A MATHEMATICAL KALEIDOSCOPE
Applications in Industry, Business and Science

Mathematics possesses not only truth, but supreme beauty - a beauty cold and austere, like that of sculpture, and capable of stern perfection such as only great art can show.

Bertrand Russell in *The Principles of Mathematics*

ABOUT OUR AUTHORS

BRIAN CONOLLY graduated with a first class honours degree in mathematics from Reading University in 1944 . He joined the Admiralty for defence work (1944-1959) where he first met and worked with his co-author Steven Vajda. His reputation in operational research was recognised by NATO for defence strategy, and he worked at their Centre for Antisubmarine Warfare at La Spezia, Italy from 1959 until 1973.

He became a professional academic, first (1967-1968) at the Virginia Polytechnic Institute and State University, Blacksburg, Virginia, USA, where he was appointed Professor of Statistics. Subsequently, from 1973 he was Professor of Mathematics (Operational Research) in the University of London, firstly at Chelsea College and afterwards at Queen Mary & Westfield College.

As a practising mathematician Brian's work has been directed towards the solution of practical problems with a particular interest is in applied probability and operational research. He has authored and co-authored some 60 papers, a large number of classified reports, and four books for the distinguished Ellis Horwood Series in Mathematics and its Applications, of which he was also a series editor.

STEVEN VAJDA is a mathematical nonagenarian, amazingly still teaching and writing in his 94th year. His contribution to operational research was recognised by the Operational Research Society, London with an award of the Companionship of Operational Research and a special issue of their Society's journal on his 91st birthday. He is a Member of the Institute of Actuaries and a Fellow of the Royal Statistical Society, London.

Steven studied mathematics in Vienna, who gave him a doctorate where after graduation (D.Phil.) he qualified as an insurance actuary and afterwards, during the Nazi regime, moved to the UK in 1939. During and after World War II he worked within the British Admiralty as a statistician and as Assistant Director of Physical Research and Operational Research, and later as Head of Mathematics Group at the Admiralty Research Laboratory, Teddington.

He was awarded awarded an honorary doctorate (D.Tech. h.c.) by Brunel University and, in 1965 became the first British Professor of Operational Research to be appointed, in Birmingham University, where he worked until retirement. In 1967 he was invited by Sussex University to become a Fellow, and in 1973 became Visiting Professor of Mathematics, in which role he continues actively, teaching and writing research papers. He has lectured in San Sebastian, Spain in tandem with Brian Conolly, his present co-author. He is a major figure in the mathematical programming community and has written 14 books, one the very first on linear programming which spread the subject throughout Europe and Japan. His books have been translated into French, German, Italian, Japanese, Russian and Spanish.

A MATHEMATICAL KALEIDOSCOPE
Applications in Industry, Business and Science

BRIAN CONOLLY, BA., MA., FSS
Emeritus Professor of Mathematics (Operational Research)
University of London

and

STEVEN VAJDA, D.Phil., D.Tech.
Visiting Professor at Sussex University
formerly Professor of Operational Research
Department of Engineering Production
University of Birmingham

Albion Publishing
Chichester

First published in 1995 by
ALBION PUBLISHING LIMITED
International Publishers, Coll House, Westergate, Chichester, West Sussex,
PO20 6QL England

British Library Cataloguing in Publication Data
A catalogue record of this book is available from the British Library

ISBN 1-898563-21-7

Printed in Great Britain by Hartnolls, Bodmin, Cornwall

Table of Contents

Preface

During our professional careers as practising and teaching mathematicians it has been a pleasant, but often frustrating experience to brush against fascinating problems on the fringe of present interests, or to realize that aspects of current work have an unexpected connection with another field. All too often, through lack of time, these have been pushed aside without follow-up. Sometimes problems have been solved but left undocumented for the same reason. Happily there is an inexhaustible supply of problems left over and curious topics worth airing. It is salutary to start each day as G.H Hardy is reputed to have done, with the intention to make attempts on outstanding unsolved problems like proving the Riemann Hypothesis, or the Goldbach Conjecture!

In this little book we offer a collection of essays derived from such sources. They are mostly quite short; "mathematical moments" we might say, but, because of the variety we call them a mathematical kaleidoscope. All are connected in some fashion with Applied Mathematics, but we leave it to the reader to decide which definition of this term would be most appropriate. The collection is aimed at a broad spectrum of mathematical readership. The material is original either in content, presentation or both. In all cases the target is the communication of our own interest in facts not easily found in modern texts, drawn from algebra, geometry, analysis, probability theory, not to mention history. We do not apologise for this. It has happened that branches of mathematics whose energy was thought exhausted undergo a renaissance which imparts renewed vigour. We have aimed to stimulate interest; some of the topics are open-ended and we hope that readers will be inspired to probe more deeply. The intention is to be instructive, and occasionally entertaining.

The collection would be a useful text for general mathematical background reading and it should not be ruled out as a companion for the bedside. It would also be a natural accompaniment for courses on the application of mathematics in the non-physical sciences, sometimes dubbed *applicable mathematics*.

1

Fantasies

Three of the five essays in this chapter (1.1,1.4,1.5) concern probability. 1.1 is an elementary application to the transmission of information through a faulty channel, followed by the evaluation of the effectiveness of an advertising campaign; 1.4 gives a light-hearted discussion of some applications of 'simple' birth-death processes, intended partly as a teaching tool, and including the spread of disease by an undetected carrier; 1.5 discusses the hypergeometric distribution with some applications.

Of the remainder, 1.3 describes a nightmare of clockface disorientation, while 1.2, in two parts, develops differential equation results paralleling those derived from the more familiar Fibonacci difference equations, discusses the peculiar uniformity characteristics that Fibonacci mechanisms bestow on pseudorandom number generators, and concludes with a glimpse of the surprising nature of ultra- (or meta-) Fibonacci mechanisms.

1.1 MISLEADING MESSAGES

1.1.1 Send Three and Fourpence, we're going to a Dance

At a military command centre on a battlefield instructions are received from headquarters over a communication link which, because of its vulnerability to enemy attack, has probability P of being faulty. **Faulty** means that a message may be garbled, that is, distorted in transmission. An example is provided by the title which was once, the legend has it, the received form of the message **'Send reinforcements, we're going to advance'**. With probability $Q(=1-P)$ the link is not defective.

For our purpose, a message consists of a sequence of N binary digits (bits), that is zeros and ones, and the question is: If an N bit message is transmitted what sort of scrambling is likely to happen to it? We are not concerned so much with the actual message content, but with its statistics and the statistics of the message received.

It is essential to specify the garbling process which will be of two kinds. The first kind is where the message is taken symbol by symbol and changed in such a way that a 0 becomes a 1, and vice versa, with probability g, and is unaltered with probability $h=1-g$. The second kind is where, if garbling operates, again with probability g, every symbol is transmuted, so that all zeros in the message become ones, and all ones become zeros. It seems, on the face of it, that the second kind will distort the meaning more than the first kind where at least some symbols will be the ones transmitted. However, our purpose is more modest. Let the number of zeros in a received message of length N be Z. This is a random variable and we shall ask, for each garbling type, what is the probability G_n that $Z=n$ and what are the means and variances. We are a long way from the actual meaning of the message, which is another topic.

Let zeros occur in the original message with probability p and ones with probability $q=1-p$. This may be interpreted to mean that we are looking at messages which at transmission contain, on average, Np zeros, though we are not concerned with the order and hence the meaning of the message.

Consider first mechanism 1. Notice that, if garbling operates, the symbol we read is a zero with probability $p'=ph+qg$ because if it **is** a zero it must not be changed, and if it is a one it must be changed. Likewise we read a one with probability

$q'=pg+qh$. Thus

$$G_{1n}=Q\binom{N}{n}p^{n}q^{N-n}+P\binom{N}{n}p'^{n}q'^{N-n}, \qquad (0\leq n\leq N).$$

A subscript has been appended to G to indicate that mechanism 1 is intended. The corresponding probability generating function $M_1(\theta)$ is

$$M_1(\theta)=Q(q+pe^{\theta})^N+P(q'+p'e^{\theta})^N,$$

from which it follows that the mean E_1 and variance V_1 are

$$E_1(Z)=N\{p+gP(q-p)\},$$

$$V_1(Z)=Npq+NPg(p-q)+N^2PQg^2(p-q)^2.$$

Note that, if $p=q$, which is equivalent to specifying that the average number of zeros and ones in the message transmitted is equal to $N/2$, possible garbling has, statistically speaking, no effect because the statistics of received messages are just those of the binomial distribution with parameters N and p.

For the second mechanism,

$$G_{2n}=Q\binom{N}{n}p^{n}q^{N-n}+P\{g\binom{N}{n}p^{N-n}q^{n}+h\binom{N}{n}p^{n}q^{N-n}\},$$

because, as before, if the channel is not defective, the number of zeros is that transmitted, while if the channel is defective and garbling obtains, the number of zeros received is the number of ones in the transmitted message, and otherwise, it is the number of zeros. The moment generating function is

$$M_2(\theta)=Pg(p+qe^{\theta})^N+(1-Pg)(q+pe^{\theta})^N.$$

The mean and variance of Z are

$$E_2(Z)=N\{p+gP(q-p)\},$$
$$V_2(Z)=Npq+N^2Pg(1-Pg)(p-q)^2.$$

It is again seen that, when $p=q$, garbling has, statistically, no effect.

The average number of zeros under both mechanisms is the same, but the variances are different. To quantify this, Table 1.1.1 shows a case where $N=100$ symbols. It is seen that when p and q are interchanged, both mechanisms produce $E(Z)=N$-expected number for original p. Also SD_2 is unchanged when p and q are interchanged, but this is not so for SD_1. These observations are easily confirmed by the formulae. We see also that SD_1 first increases with P and then decreases.

This happens also with SD_2 when g increases sufficiently. See Table 1.1.2.

Table 1.1.1 Mean and Standard Deviation of Number of Zeros received at Output of a possible faulty Communication Channel

(i) Expected number of zeros in original message $=10$;
conditional garbling probability $g=0.1$

	Channel reliability P	E(Z)	$SD_1(Z)$	$SD_2(Z)$
High	0.1	10.8	3.74	8.51
Medium	0.5	14	4.58	17.69
Low	0.9	17.2	2.75	23.09

(ii) Expected number of zeros in original message $=90$;
conditional garbling probability $g=0.1$

	Channel reliability P	E(Z)	$SD_1(Z)$	$SD_2(Z)$
High	0.1	89.2	3.94	8.51
Medium	0.5	86	5.39	17.69
Low	0.9	82.8	4.69	23.09

Table 1.1.2 (i) Expected number of zeros in original message =10; conditional garbling probability g=0.9

	Channel Reliability P	E(Z)	$SD_1(Z)$	$SD_2(Z)$
High	0.1	17.2	21.64	23.09
Medium	0.5	46	35.62	39.91
Low	0.9	74.8	20.27	31.53

(ii) Expected number of zeros in original message = 90; conditional garbling probability g=0.9

	Channel Reliability P	E(Z)	$SD_1(Z)$	$SD_2(Z)$
High	0.1	82.8	21.97	23.09
Medium	0.5	54	36.62	39.91
Low	0.9	25.2	23.25	31.53

The figures confirm initial expectation that the second mechanism, where a garbled message has all symbols reversed, induces a larger variation in the number of zeros transmitted than the first mechanism where some symbols, at least,remain unchanged.

"That is all very well", announced a voice from behind my back, belonging to a military-looking gentleman who had,seemingly, been reading what I wrote over my shoulder, "but what about the distortion of the meaning of the message?" It crossed my mind to tell him to mind his own business, but as he had a point I thought I might as well say something on the subject. "If you had paid attention to what I have been writing", I replied severely, "you would know that I was not going to discuss specialised matters like the transmission of information, but since you mention it I will make a small concession."

One way of measuring the distortion of a message transmitted over a "noisy channel", as it is sometimes called, is to introduce a **discrepancy** D defined to be the sum of the number of all the inverted symbols. If this is small, language redundancy and judicious use of error-correcting codes may help to restore the meaning.For example, redundancy helps a receiver to decode fairly easily the message:'Seal zooks dpbste is piotnless. Whan woll thp rial baoks jrgpments end?'. Here there are 57 symbols with 16 letter replacements, but, especially if one knows the context, it is not hard to sort out the proper message.[1]

Returning to our message consisting of a sequence of N zeros and ones, we note that under the first garbling mechanism a faulty channel implies that the probability that

[1]Real books debate is pointless. When will the real books argument end?

n mutations occur in a particular message is binomial with success probability g. Thus,

$$P(D=0)=Q+Pg^N, \quad P(D=n)=P\binom{N}{n}g^n g^{N-n}, \quad (0<n\leq N).$$

The smaller n, the greater the chance that redundancy will allow the true meaning to be guessed/read, so we are interested in $P(D<N_R)$, where N_R is an agreed threshold for redundancy to help. To see how this might work let us take the short message 'ADVANCING'. In Morse code, with 0 for dot and 1 for dash, this is:
01,100,0001,01,10,1010,00,10,110/
so $N=24$. For the example let us suppose that $N_R=fN$ with $f=0.2$, and also take $P=0.2$, $g=0.5$, so that if the channel is faulty there is a 50-50 chance that a symbol is inverted. Now $fN=4.8$ and

$$P(D<4.8)=0.8+0.2\sum_{n=0}^{4}\binom{24}{n}(0.5)^{24}$$

$$=0.80015439.$$

This seems to give a reasonable chance that the meaning of the message will be transmitted. One crude way of improving on it is to repeat the transmission. In this case, a single repetition would yield a probability of $1\text{-}(1\text{-}0.8002)^2=0.96\ldots$ that at least one of the repetitions satisfies the criterion. This discussion would lead naturally, if we had the space and time, to the fascinating topic of error-correcting codes, useful in the reconstruction of dirty pictures and defective compact discs, to give two examples.

The voice over my shoulder made a grunting sound which I took to express thanks, but when I turned round I saw that he had turned into a pig!

1.1.2 The advertising campaign

"How",said my friend, the impoverished inventor, "can I establish a market for my new mousetrap? It works on a novel principle, is humane, elegantly designed, inconspicuous, and inexpensive to manufacture. No household will tolerate existence without at least one, and if I can only let the world know about it, I shall be on to a winner. Whatever can I do?". My friend is ingenious and resourceful, but he lives in another world, remote from reality, filled with gadgets for removing stones from hooves, unbreakable violin strings, self-sharpening razor-blades, and a multitude of other devices of potential benefit to humanity, if only humanity could be made aware.

"What you need", I told him, "is a well-orchestrated advertising campaign. Television is rather expensive, but you could try the newspapers. Of course", I continued,"there will be a lot more to do than just advertise. If a large demand is stimulated" (I smiled to myself) "you will have to be ready to meet it and, in any case you will have to produce a substantial initial stockpile. You will also need storage and delivery capacity. It might pay you to read my little essay on inventory and stock control."

"Bother that," said my unworldly friend, "you are always such a wet blanket. Can't you just advise me first about the advertising? How much will it cost?"

So, I went home, sat at my desk, took pen and paper, and made a mathematical model. Let p be the probability that a potential customer is sufficiently influenced by an advertisement to decide to buy the product. The potential customer is, in a sense, a target, and the advertiser is the searcher who has probability p of detecting (persuading the target to buy) whenever that target is 'illuminated' by an advertisement. The success of the advertiser at some time during the first n attempt/advertisement(s) therefore has probability $p+qp+q^2p+\ldots+q^{n-1}p=1-q^n$. Suppose that there is a pool of N potential customers and that n advertisements are placed, let us say on successive days, though this is not important. The probability $v_r^{(n)}$ that r customers will be sufficiently motivated to buy the product is binomial with parameters N and $(1-q^n)$. Thus

$$v_r^{(n)}=\binom{N}{r}(1-q^{\,n})^r(q^{\,n})^{N-r}.$$

This is not quite enough, for the advertiser really wants to know the value of the probability u_n that it takes exactly n advertisements to convert all N potential customers to the product. The formula is

$$u_n=(1-q^{\,n})^N-(1-q^{\,n-1})^N$$

because this is the difference between the probability that they have all been converted by the time of the n-th advertisement and the probability that they have been converted by the $(n$-1)-th. (Note also that

$$u_n=\sum_{r=0}^{N-1}v_r^{(n-1)}p^{N-r},$$

since either n-1 advertisements convert 0 customers and N are converted by the n-th, or 1 customer is persuaded by n-1 advertisements and N-1 by the n-th,...,N by n-1 advertisements and one by the n-th.) Finally, the advertiser wants the probability w_n that all N potential customers targeted have been convinced to buy during the campaign of n days of advertising, and that is

$$w_n=\sum_{r=1}^{n}u_r=(1-q^{\,n})^N.$$

Let us look at some values and assess the publicity costs that the inventor may have to incur. He decides to make a modest beginning by attempting to sell 5000 mousetraps.

A local newspaper would therefore be an appropriate as well as cheaper medium. He consults a publicity agent friend who makes an eyecatching design and advises that a half-page would be a suitable space to rent. The agent thinks, somewhat immodestly, that few will be able to resist the allure of his design and that p should be assessed at 0.9. I, however, on being asked my opinion, suggest the more conservative estimate of 0.2. We decide to make calculations for both assumptions to see how u_n grows with n. The results are shown in Table 1.1.2.

Table 1.1.2 Probability that a Campaign of *n* Advertisements will convince *N* Customers to buy Product

Persuasive power of advert represented by probability p

N	5000	2500	1000	5000	2500	1000	5000	2500	1000
p	0.2			0.5			0.9		
n	w_n	w_n	w_n	w_n	w_n	w_n	w_n	w_n	w_n
1	0	0	0	0	0	0	0	0	0
2	0	0	0	0	0	0	0	0	0
3	0	0	0	0	0	0	.007	.082	.368
4	0	0	0	0	0	0	.606	.779	.905
5	0	0	0	0	0	0	.951	.975	.990
10	0	0	0	.008	.087	.376	1	1	1
15	0	0	0	.858	.926	.970			
20	0	0	0	.995	.998	.999			
30	.002	.045	.290	1	1	1			
40	.514	.717	.876						
45	.804	.897	.957						
50	.931	.965	.986						

If the advertising consultant had been as effective as he claimed, a campaign of 5 days would have 95% probability of capturing 5000 customers, 97% of capturing 2500, and 99% for 1000. Even with the more moderate estimate of $p=0.5$, a 20-day campaign would have a very high success rate, but what I thought to be the more realistic estimate, $p=0.2$, would require a protracted campaign and, probably, a heavy outlay on advertising, even in a provincial newspaper.

1.2 FIBONACCI FANTASIES

1.2.1 Mathematicians are simple folk, easily amused, quiet and amiable. They have no need for extraneous diversions since there are plenty of inner thoughts and puzzles to keep the mind busy. They make good husbands, even if their abstraction is sometimes infuriating. And lest I be thought sexist I should add that they also make good wives when they are of the feminine gender.

One fertile topic, tempting and not excessively demanding of a mathematician's resources, was furnished originally by a certain Signor Fibonacci. "Fertile" is a just word since what has grown into nothing short of a Fibonacci industry started as an exercise in applying mathematics to the procreation of rabbits. If we play with the fecundity of the rabbits, or transfer the model to the growth of other biomasses, malignant cells, for example, we are led to the study of properties of solutions of the second order, linear difference equation

$$x_{n+2} = a\,x_{n+1} + b\,x_n\,, \tag{1.2.1}$$

supplemented by appropriate starting values x_0 and x_1. a and b can be more or less anything, but we mostly think of them as integers. It is more interesting if $a^2+4b\neq 0$, and the original Fibonacci case has $a=b=1$. For an up-to-date account of Fibonacci sequences, their properties and related topics, see [1]. It will help the reader to have a little general preliminary theory in the text for reference. We propose to sketch more detailed proofs of the results to follow in the Notes.

Let G_n be a general solution of (1.2.1). If α and β are zeros of t^2-at-b with, if necessary, the convention that $|\alpha| \geq |\beta|$, general theory prescribes that any solution of (1.2.1) can be written as a linear combination of powers of α and β. Thus,

$$G_n = A\alpha^n + B\beta^n, \tag{1.2.2}$$

and, matching the constants A, B to the initial values G_0, G_1 permits the form

$$G_n = \frac{(G_1 - G_0\beta)\alpha^n + (G_0\alpha - G_1)\beta^n}{\alpha - \beta}. \tag{1.2.3}$$

α and β are given, respectively, by

$$\alpha = (a+r)/2, \qquad \beta = (a-r)/2, \qquad r^2 = a^2+4b$$

with the properties

$$\alpha + \beta = a, \qquad \alpha\beta = -b. \tag{1.2.4}$$

Thus, (1.2.3) can be written

$$G_n = G_1 F_n + bG_0F_{n-1}, \qquad (1.2.5)$$

where F_n is the "standard" solution of (1.2.1) with $F_0=0$, $F_1=1$, giving

$$F_n = \frac{\alpha^n - \beta^n}{\alpha - \beta}. \qquad (1.2.6)$$

(1.2.6) holds for positive and negative integer n, and

$$F_{-n} = (-)^{n+1} F_n / b^n, \qquad (1.2.7)$$

which means that when $b>1$, (F_n) has fractional values for negative n and alternating signs.
It is convenient to introduce a second, independent standard solution L_n of (1.2.1) given by

$$L_n = \alpha^n + \beta^n \qquad (1.2.8)$$

so that $L_0=2$, $L_1=a$.

When $a=b=1$, (L_n) is the *Lucas* sequence. Justified by the general theory we can express G_n as a linear combination of F_n and L_n.
Thus, matching initial conditions, we have

$$G_n = (G_1 - aG_0/2)F_n + G_0L_n/2. \qquad (1.2.9)$$

The first of our fantasies will be to make a selection from the list of Fibonacci formulae on pp.176-184 of [1] and to develop their counterparts for general a and b. But first, here is a list of values of F_n and L_n for $n=1..7$.

n	F_n	L_n
0	0	2
1	1	a
2	a	a^2+2b
3	a^2+b	a^3+3ab
4	a^3+2ab	$a^4+4a^2b+2b^2$
5	$a^4+3a^2b+b^2$	$a^5+5a^3b+5ab^2$
6	$a^5+4a^3b+3ab^2$	$a^6+6a^4b+9a^2b^2+2b^3$
7	$a^6+5a^4b+6a^2b^2+b^3$	$a^7+7a^5b+14a^3b^2+7ab^3$

It is sometimes a surprise to be told that all true F_n whose indices are divisible by 5 are themselves divisible by 5. In the general case it is easily checked that F_{10} is divisible exactly by F_5, and a little thought explains the reason: α^{10}-β^{10} is divisible by α^5-β^5, and so is α^{5n}-β^{5n}. Similarly all even order F_n are divisible by $F_2=a$, and the imagination needs little stimulation to see where that particular path leads.

But now for some specific relations. The numbers refer to the list mentioned above.

(1) (Definition) $F_{n+2}=aF_{n+1}+bF_n$

(2) $F_{-n}=(-)^{n+1}F_n/b^n$ {see (1.2.7) above}

(4) $L_{-n}=(-)^nL_n/b^n$

(5) $bL_{n-1}+L_{n+1}=r^2F_n$

(6) $F_{n+1}+bF_{n-1}=L_n$

(8) $G_{m+n}=bF_{m-1}G_n+F_mG_{n+1}$

(11) $F_{n+1}^2+bF_n^2=F_{2n+1}$, and

$$F_{n+1}^2-bF_n^2=aF_{n+2}F_{n-1}+(-b)^{n-1}(a^2-b)$$

(18) Let G_n, H_n be arbitrary general solutions, represented by

$$G_n=A\alpha^n+B\beta^n, \quad H_n=P\alpha^n+Q\beta^n,$$

where A,B,P,Q are arbitrary constants (i.e. n-independent and determined by the initial conditions assigned). Then

$$G_{n+h}H_{n+k}-G_nH_{n+h+k}=(-b)^n(G_hH_k-G_0H_{h+k}).$$

(46) The following is a binomial coefficient result:

$$G_{m+n}=\sum_{i=0}^{m}\binom{m}{i}a^{m-i}b^iG_{n-i}.$$

(106) We jump to the end, a finite continued fraction.

$$\frac{F_{(t+1)m}}{F_{tm}} = L_m - \frac{(-b)^m}{L_m -}\frac{(-b)^m}{L_m -}\cdots\frac{(-b)^m}{L_m}.$$

This finite continued fraction has t terms.

And, finally, two series not mentioned in [1].

$$\frac{-\beta}{\alpha} = \sum_{n\geq 1} \frac{(b/r)^n \beta^{n(n-1)/2}}{F_1 F_2 \ldots F_{n+1}}. \tag{1.2.10}$$

Next, let

$$S = \sum_{n\geq 1} \frac{1}{F_n}. \tag{1.2.11}$$

Then another representation is

$$S = 1 + \sum_{n\geq 1} x_1 x_2 \ldots x_n,$$

where x_n is the *n-th* convergent of the continued fraction representation of

$$\alpha - a = (r-a)/2 = -\beta.$$

Sketches of proofs are given in the Notes.

We now introduce a species of **Fibonacci function.** Such functions depend on a real or complex variable x, and to G_n, F_n, L_n, etc., are to correspond $G(x)$, $F(x)$, $L(x)$,... A possible defining mechanism is

$$G(x) = \sum_{n\geq 0} G_n \frac{x^n}{n!}. \tag{1.2.12}$$

Since

$$G_n = A\alpha^n + B\beta^n,$$

this definition leads to the representation

$$G(x) = Ae^{\alpha x} + Be^{\beta x}. \tag{1.2.13}$$

Returning to the definition (1.2.12), and denoting differentiation with respect to argument x by the operator D, we find

$$DG(x)=\sum_{n\geq 0} x^{n-1}G_n/(n-1)!=\sum_{n\geq 0} x^n G_{n+1}/n!$$

and

$$D^2G(x)=\sum_{n\geq 0} x^{n-2}G_n/(n-2)!=\sum_{n\geq 0} x^n G_{n+2}/n!, \qquad (1.2.14)$$

from which we conclude that $G(x)$ is a solution of a second-order, linear differential equation, namely

$$(D^2-aD-b)G(x)=\sum_{n\geq 0} x^n(G_{n+2}-aG_{n+1}-bG_n)/n!=0, \qquad (1.2.15)$$

since G_n satisfies (1.2.1). This is consistent with (1.2.13), and either (1.2.15) or (1.2.12) could be adopted as definition of $G(x)$. As the summations are from $n=0$, we have $G(0)=G_0$ and $G'(0)=G_1$, where primes denote differentiation, and thus G(x) can be written as

$$G(x)=Ae^{\alpha x}+Be^{\beta x}=\frac{(G_1-G_0\beta)e^{\alpha x}+(G_0\alpha-G_1)e^{\beta x}}{\alpha-\beta}. \qquad (1.2.16)$$

We can now mimic the earlier introduction of F_n and L_n by presenting $F(x)$ and $L(x)$ as

$$F(x)=\frac{e^{\alpha x}-e^{\beta x}}{\alpha-\beta},$$

$$L(x)=e^{\alpha x}+e^{\beta x} \qquad (1.2.17)$$

and then

$$G(x)=G_1F(x)+bG_0D^{-1}F(x), \qquad (1.2.18)$$

where D^{-1} is to be interpreted as the integral operator without the constant, thus

$$D^{-1}e^{\alpha x} = e^{\alpha x}/\alpha.$$

Comparison with (1.2.5) suggests the beginning of a formal correspondence between solutions of difference equations of Fibonacci family and their functional counterparts defined here, namely

$$G_n \to G(x), \qquad G_{n-1} \to D^{-1}G(x)$$

and, as we shall see,

$$G_{n+1} \to DG(x). \tag{1.2.19}$$

Before pursuing this, we note finally that

$$G(x) = (G_1 - aG_0/2)\ F(x) + G_0\ L(x)/2, \tag{1.2.20}$$

the counterpart of (1.2.9).
Let

$$G(x) = Ae^{\alpha x} + Be^{\beta x}, \quad H(x) = Pe^{\alpha x} + Qe^{\beta x}, \tag{1.2.21}$$

be two independent solutions of the differential equation

$$(D^2 - aD - b)y = 0.$$

The Wronskian $W(x)$ of these solutions is given by

$$W(x) = G(x)H'(x) - G'(x)H(x) = (\alpha - \beta)e^{ax}(BP - AQ), \tag{1.2.22}$$

for which a little algebra is needed. The discrete counterpart should involve $G_nH_{n+1} - H_nG_{n+1}$, and with the definition in (2.18) of G_n, H_n we obtain

$$G_nH_{n+1} - H_nG_{n+1} = (-b)^n(\alpha - \beta)(BP - AQ). \tag{1.2.23}$$

Thus (1.2.22) and (1.2.23) are counterparts. Note the special case of (1.2.22),

$$F(x)L'(x) - F'(x)L(x) = -2e^{ax},$$

and its counterpart

$$F_nL_{n+1} - F_{n+1}L_n = -2(-b)^n. \tag{1.2.24}$$

(1.2.23) and (1.2.24) are special cases of [1],(18). These are some discrete formulae with simple subscripts viewed against their counterparts. To conclude we investigate the more complicated G_{m+n} and its functional counterpart. [1],(8), p.176, is for convenience repeated:

$$bF_{m-1}G_n + F_mG_{n+1} = G_{m+n}. \tag{1.2.25}$$

To find its counterpart, consider

$$S = \sum_{m \geq 0} \frac{y^m}{m!} \sum_{n \geq 0} \frac{x^n}{n!} (bF_{m-1}G_n + F_mG_{n+1}). \tag{1.2.26}$$

Using the formalism (1.2.12) gives

$$S=bG(x)\sum_{m\geq 0}\frac{y^m}{m!}F_{m-1}+F(y)\sum_{n\geq 0}\frac{x^n}{n!}G_{n+1}$$

$$=bG(x)\sum_{m\geq 0}\frac{y^{m+1}}{(m+1)!}F_m+F(y)\sum_{n\geq 0}\frac{x^{n-1}}{(n-1)!}G_n$$

$$=bG(x)D^{-1}F(y)+F(y)DG(x),$$

and, using (1.2.13) and (1.2.17), we can reduce the right hand side to give

$$S=Ae^{\alpha(x+y)}+Be^{\beta(x+y)}=G(x+y)$$

so that

$$bG(x)\mathrm{D}^{-1}F(y)\ +\ F(y)\mathrm{D}G(x)\ =\ G(x+y). \tag{1.2.27}$$

But also, substituting (1.2.25) in the right hand side of (1.2.26), gives

$$S=\sum_{m\geq 0}\frac{y^m}{m!}\sum_{n\geq 0}\frac{x^n}{n!}G_{m+n}, \tag{1.2.28}$$

and this could be argued to represent $G(x+y)$ in the same sense that the right hand side of (1.2.12) 'is' $G(x)$. The interesting feature here is the correspondence that would enable functional and difference equations to be written down, one from the other, using simple rules. Encouraged by these findings we offer the reader the following companion to [1], (18):

$$\frac{G(x+y)H(x+z)-G(x)H(x+y+z)}{G(y)H(z)-G(0)H(y+z)}=e^{ax}. \tag{1.2.29}$$

1.2.2 In this part we shall examine the capability of Fibonacci-type mechanisms to generate **uniform digits**. This topic is of interest to scientists concerned with the use of computers for **simulation**. This is an experimental technique much in vogue in operational research and statistics. The idea is to program on a digital computer a representation of a complex random process that can not be dealt with satisfactorily by analysis and to observe the effect on the output of variations in the parameters of the process. At the heart of this technique is the requirement to generate samples from prescribed probability distributions, and to do this necessitates the production of streams of real numbers which, if not truly uniformly distributed over (0,1), at least give the appearance of being so. Such sequences are called *pseudouniform*.

The most common mechanisms for this are difference equations, and in this family the generalised Fibonacci mechanism plays an interesting role. Thus we consider here sequences (u_n) generated by

$$u_{n+2} \equiv au_{n+1} + bu_n \bmod p, \text{ (} a \text{ and } b \text{ are positive integers)} \tag{1.2.30}$$

with $u_0=0$, $u_1=1$, and where p is an odd prime of the form

$$p=a^2+4b, \ (a>0, b>0) \tag{1.2.31}$$

so that a is odd. The development and results which follow depend on a little number theory summarised in the Notes.

We shall show that such sequences have period $p(p-1)$ (that is they return to the initial pair (0,1) for the first time when $p(p-1)$ numbers have been produced) and consist precisely of $p-1$ of each of the least residues $0,1,2,\ldots,p-1$. Thus the mechanism does indeed yield a kind of guaranteed uniform distribution of integers which, after division by p, belong to the interval (0,1). This is already well-known for $p=5$, corresponding to the classical Fibonacci mechanism with $a=b=1$. For background and references see [1], Chapter VII. For illustration in a form later to be generalised the sequence is:

n	u_n	n	u_n	n	u_n	n	u_n
0	0	5	0	10	0	15	0
1	1	6	3	11	4	16	2
2	1	7	3	12	4	17	2
3	2	8	1	13	3	18	4
4	3	9	4	14	2	19	1

The reader will readily check that indeed 4 of each of the least residues mod 5, viz. 0,1,2,3,4, appear. In the general case,

$$u_{n+2}=au_{n+1}+bu_n,$$

and the solution (1.2.6) with $u_0=0$, $u_1=1$, is

$$u_n=\frac{\alpha^n-\beta^n}{\alpha-\beta}$$

$$\text{with } \alpha=(a+r)/2, \ \beta=(a-r)/2, \ r^2=a^2+4b.$$

This gives, on expansion of α and β,

$$u_n=\frac{1}{2^{n-1}}\left[\binom{n}{1}a^{n-1}+\binom{n}{3}a^{n-3}r^2+\ldots\right]\equiv n\left(\frac{a}{2}\right)^{n-1} \bmod p, \tag{1.2.32}$$

since $r^2=p$.

Let

$$\tfrac{1}{2}a\equiv c \bmod p.$$

Then

$$u_n\equiv nc^{n-1} \bmod p. \tag{1.2.33}$$

Now construct an array with p-1 vertical divisions and p rows. The vertical divisions contain three columns each, the first for n, the second for c^{n-1}, and the third for the product nc^{n-1}, all reduced modulo p. This is shown in the following table.

Table 1.2.1 **Values of $u_n \equiv nc^{n-1}$ mod p**

n	c^{n-1}	u_n	n	c^{n-1}	u_n	n	c^{n-1}	u_n		n	c^{n-1}	u_n
0	c^{p-2}	0	$p\equiv 0$	$c^{p-1}\equiv 1$	0	$2p\equiv 0$	$c^{2p-1}\equiv c$	0	...	0	c^{p-3}	0
1	$c^{p-1}\equiv 1$	1	$p+1\equiv 1$	c	c	$2p+1\equiv 1$	c^2	c^2	...	1	c^{p-2}	c^{p-2}
2	c	$2c$	$p+2\equiv 2$	c^2	$2c^2$	$2p+2\equiv 2$	c^3	$2c^3$	...	2	$c^{p-1}\equiv 1$	$2c^{p-1}$
3	c^2	$3c^2$	$p+3\equiv 3$	c^3	$3c^3$	3	c^4	$3c^4$	...	3	c	$3c$
...	...	..	...	...	..	...	...	...	...	..	...	...
$p-2$	c^{p-3}	$(p-2)c^{p-3}$	$p-2$	c^{p-2}	$(p-2)c^{p-2}$	$p-2$	$c^{3p-3}\equiv 1$	$p-2$	...	$p-2$	c^{p-4}	$(p-2)c^{p-4}$
$p-1$	c^{p-2}	$(p-1)c^{p-2}$	$p-1$	1	$p-1$	$p-1$	$c^{3p-2}\equiv c$	$(p-1)c$	...	$p-1$	c^{p-3}	$(p-1)c^{p-3}$

We look first at the first triplet of columns. n runs downwards from 0 to p-1 and c^{n-1} from $c^{p-2} \equiv c^{-1}$ (since $c^{p-1} \equiv 1$) through $c^{p-1}, c, c^2, \ldots,$ to c^{p-2} again. Next look at the second triplet. n runs from p to $2p$-1, or from 0 to p-1 mod p so that n has, mod p, the same value in each row as it had in the first division. But the c^{n-1} column starts at $c^{p-1} \equiv 1$, and as we run down the column we see that in each place c^{n-1} has index numerically 1 more than in the previous division at the same location. Similar considerations apply to the remaining divisions. Modulo p, n has always the same value at each row as in previous divisions while c^{n-1} has index 1 more than in the previous division. Running along the table sideways, in each row we observe that the entries in the c^{n-1} columns contain one each of $c, c, c^2, \ldots, c^{p-1}$ in ascending order of the index.

Case 1: c is a primitive root of p

If c is a primitive root of p this means that the sequence of c^{n-1} entries in each row constitutes a permutation of the set of least positive residues mod p. Multiplication in each row of c^{n-1} by the corresponding n yields another permutation of the set of least positive residues and, as a consequence, the u_n columns must altogether contain p-1 of each of the p least residues modulo p, namely $0, 1, \ldots, p$-1. This is exemplified for $p=5$ by the table given earlier.

Case 2: c is not a primitive root of p

If c is not a primitive root of p, but has cycle length m, say (a submultiple of p-1), consisting of a set S of elements $(s_1 = c, s_2 = c^2, \ldots, s_m = c^m \equiv 1)$, it is clear that each row of the table contains k repetitions of the elements of S, where $k = (p\text{-}1)/m$. Reference to the Notes should persuade the reader that, taken over the whole table, to each single appearance of S correspond in the nc^{n-1} columns m copies of each of the members of R^+. Since there are k appearances of S in each row, this means that each member of R^+ appears $km=p-1$ times. This is the basis for the assertion that the "Fibonacci method" can be made to yield a kind of uniform distribution of reals over (0,1). It has been shown (see[1]) in the case $a=b=1$ and hence $p=5$ that a uniform distribution mod p^n can be obtained by the same method.

A problem for the reader is: Does this result generalize?

1.2.3 We now know that the difference equation, written with the subscripts as arguments,

$$u(n)=u(n-1)+u(n-2), \quad u(0)=0, \quad u(1)=1, \qquad (1.2.34)$$

yields a sequence of integers (both backwards and forwards) with many interesting properties and applications. An apparently related path which invites exploration is opened if the arguments in (1.2.34)) are made to depend on previous members of the sequence. An example, introduced by Hofstadter [3], is the mechanism

$$H(1)=H(2)=1, \quad H(n)=H(n-H(n-1))+H(n-H(n-2)), \quad (n>2). \qquad (1.2.35)$$

If it were possible to perform genetic engineering on the Fibonacci mechanism it is not hard to imagine that something like (1.2.35) could emerge. The first 64 values of $H(n)$ (and also of some other close relatives) are printed in Table 1.2.2. These, to some extent, explain Hofstadter's comment that the $H(n)$ sequence is "chaotic" and that it leads to "a small mystery...the further out you go...the less sense it seems to make...". This is borne out by Table 1.2.2. **'Chaos'** is a popular topic at the time of writing. You start with an apparently well-behaved mechanism, make some minute perturbation - in a parameter, or in the initial values - and all hell breaks loose. The H mechanism does not quite conform to this description but what we see by comparison with the other sequences in Table 1.2.2 is a mechanism struggling to escape from a straitjacket. Perhaps it does - eventually. That is one aspect of this puzzle.

The sequences in Table 1.2.2 labelled $F(n)$, $C(n)$ and $N(n)$ are defined as follows for $n>2$:

$$F(1)=0 \text{ or } 1, \; F(2)=1, \; F(3)=F(n-F(n-1))+F(n-1-F(n-2)), \qquad (1.2.36)$$

$$C(1)=C(2)=1, \; C(n)=C(n-C(n-1))+C(C(n-1)), \qquad (1.2.37)$$

$$N(0)=N(1)=N(2)=1, \; N(n)=N(n-1-N(n-1))+N(n-2-N(n-2)) \; . \qquad (1.2.38)$$

These are all of a family with H and are reminiscent of the mechanism (1.2.34), perhaps (1.2.36) and (1.2.38) more than the others. They differ from H in that closed form

representations have been found. Yet another cousin is
$K(1)=K(2)=1,\ K(n)=K(K(n-1))+K(K(n-2)),\ (n>2).$ (1.2.39)
From n=3 onwards the sequence so defined has the constant value 2.

Table 1.2.2 (page 29) shows that these sequences behave with exemplary docility, at least as compared with H, yet their mechanisms are similar. They appear to be monotonic, taking every integer value from 1 to 32, with some repetition, whereas H can be seen to miss 7,13,15,18,27,29. It also misses 34,49,51,59,..., and, no doubt, many more as n progresses. A more extensive Table is given in [1].

It seems that the progression of F, C, N is naturally divisible into cycles, or perhaps a better terminology would be *octaves*, beginning at $n=2^i$ and ending at $2^{i+1}-1$ for $i=1,2,\ldots$. It appears, moreover, that F, C, N all take the value 2^{i-1} at $n=2^i$. Some values taken by H at $n=2^i$ are shown in Table 1.2.3; they are more ragged. It is shown in [1] that for $n\geq 2$, and m the largest integer such that $n=2^m+k$, $(m\geq 1,\ 0\leq k\leq 2^m-1)$ (i.e. in the m-th octave)

$$F(n)=2^{m-1}+F(k+1),(n\geq 2).$$

This requires the definition $F(1)=0$ for consistency BUT THE SEQUENCE IS THE SAME WHETHER $F(1)=0$ or 1!

The behaviour of $C(n)$ is more complex. The mechanism itself was apparently proposed by J. Conway who stung Mallows into action at a lecture at the Bell Labs. (see [4]). A sufficient analysis was however published previously in [1] to untangle the structure of the sequence and to define an algorithm by which any $C(n)$ can be obtained.

The present writer,having wrestled intermittently for some time, primarily with H and then with C, dealing with F on the way (inspired by [5] and correspondence with Professor Guy), proposed $N(n)$ as positively his final variant on the meta-Fibonacci theme. The solution can be written

$$N(2^m+k)=2^{m-2}+N(2^{m-1}+k-1), \quad (k=1,2,\ldots,2^{m-1}+1);$$

$$N(2^m+2^{m-1}+k)=2^{m-1}+N(2^{m-1}+k), \quad (k=0,1,\ldots,2^{m-1}).$$

This sequence has also been studied by Professor Stephen Tanny at the University of Toronto. The discussion leaves a lot of scope for the 'interested reader'. I recommend also that such a person read [4] by way of an introduction.

1.2.4 Notes

1.2.4.1 The following notes refer to the generalized forms in the first part of the text.

(2) This, and all the other results, can be deduced from the representation of the elements in terms of the roots α and β of $t^2-at-b=0$. Here the definition of F_{-n} is

$$(\alpha-\beta)F_{-n}=\alpha^{-n}-\beta^{-n}=(\beta^n-\alpha^n)/(\alpha\beta)^n=-(\alpha-\beta)F_n/(-b)^n,$$

from which the result follows.

(4) Similar.

(5) $bL_{n-1}+L_{n+1}=bL_{n-1}+aL_n+bL_{n-1}=(\alpha+\beta)(\alpha^n+\beta_n)-2(\alpha\beta)(\alpha^{n-1}+\beta^{n-1})$

$$=(\alpha-\beta)(\alpha^n-\beta^n)=(\alpha-\beta)^2F_n=r^2F_n.$$

(6) Similar.

(8) Follows from α,β definitions.

(11) Show that $F_{n+1}^2+bF_n^2-F_{2n+1}=0$. It is convenient to isolate the α^{2n} and β^{2n} terms.

The second formula is similar.

(18) Use the definitions to remove the factor $(\alpha\beta)^n$ from the left side.

(106) It is easy for $t=1$, starting from the definitions of F, i.e. $F_{2m}/F_m=(\alpha^{2m}-\beta^{2m})/(\alpha^m-\beta^m)=\alpha^m+\beta^m=L_m$.

In general show that

$$\frac{F_{(t+1)m}}{F_{tm}}=L_m-\frac{(-b)^m}{F_{tm}/F_{(t-1)m}},$$

and the result follows by continuing the process, but only t times.

The series defined in (1.2.10) starts from the identity

$$1=\frac{1}{1-x}-\frac{x}{1-x}[\frac{1}{1-x^2}-\frac{x^2}{1-x^2}[\frac{1}{1-x^3}etc.........]]...].$$

Now put $x=\beta/\alpha$, and the best of luck!

The series identified as (1.2.11) can be written

$$S=\frac{1}{F_1}+\frac{1}{F_1}\frac{F_1}{F_2}+\frac{1}{F_1}\frac{F_1}{F_2}\frac{F_2}{F_3}+...$$

and with $F_1=1$ and $x_n=F_n/F_{n+1}$ the form

$$S=1+x_1+x_1x_2+x_1x_2x_3+...$$

is obvious. Now if $x=\alpha-a=(r-a)/2=2b/(r+a)=b/(a+x)$, the continued

fraction form

$$x=\frac{b}{a+}\frac{b}{a+}\frac{b}{a+}.....$$

follows. The n-th convergent to x, written as $x_n=a_n/b_n$ satisfies the relations $a_n=a_{n-1}+ba_{n-2}$, $b_n=ab_{n-1}+ab_{n-2}$, with a_0 and b_0 defined as 0 and 1, respectively. This is standard continued fraction theory. It follows that $a_n=F_n$, $b_n=F_{n+1}$, and so does the result. This is not particularly beautiful, but the continued fraction form offers computational advantage.

1.2.4.2. Number Theoretic Complements

In these notes and the related part of the text we deal exclusively with integers, so roman type will be used. A positive integer p which can be divided without remainder only by itself and by 1 is called a *prime*. Apart from p=2, all primes are odd integers. The set of integers R=(0,1,...,p-1) constitutes the *set of least residues modulo p*. Note that "modulo" is usually abbreviated to "mod". Excluding 0 from R leaves the set R^+ of least positive residues mod p. All integers n, positive and negative, including zero, have a unique correspondence with a member of R. The algorithm for finding that member r(n) of R corresponding to n is r(n)=n-p[n/p], where [x]=least integer which does not exceed x. It is then customary to write n≡r(n) mod p, which is said 'n is congruent to r(n) mod p'. A rearrangement of R from its natural order is called a *permutation of R*. Multiplication of R by any of its members other than zero results in a permutation. For consider r_1r_2, where r_1 and r_2 are members of R^+. If $r_1r_2\equiv r_1$ mod p, then $r_1(r_2-1)\equiv 0$ mod p, which means that $r_2\equiv 1$. Otherwise $r_1r_2\equiv r_3$, different from either r_1 and r_2. Thus, to multiply each member of R^+ by any integer c belonging to R^+ yields a permutation of R^+.It follows that if $cr_1\equiv s_1$, $cr_2\equiv s_2,\ldots,cr_{p-1}\equiv s_{p-1}$, then since $r_1r_2\ldots r_{p-1}=(p-1)!$, multiplication gives $c^{p-1}(p-1)!\equiv(p-1)!$ mod p, or $c^{p-1}\equiv 1$ mod p. Of course, c^n may be congruent to 1 for powers n less than p, but they must clearly be submultiples of p-1. When n=p-1 is the smallest integer for which $c^n\equiv 1$ mod p, c is said to be a *primitive root* of p. For example, consider p=13. There follows a table of c^n mod 13 for c=2(1)12.

Table of c^n

c\n	1	2	3	4	5	6	7	8	9	10	11	12
2	2	4	8	3	6	12	11	9	5	10	7	1
3	3	9	1									
4	4	3	12	9	10	1						
5	5	12	8	1								
6	6	10	8	9	2	12	7	3	5	4	11	1
7	7	10	5	9	11	12	6	3	8	4	2	1
8	8	12	5	1								
9	9	3	1									
10	10	9	12	3	4	1						
11	11	4	5	3	7	12	2	9	8	10	6	1
12	12	1										

This table shows that for $p=13$, the primitive roots are 2,6,7,11. 3 and 9 *have period 3*, and *cycle* (3,9,1) or its permutation (9,3,1), 5 and 8 have period 4, 4 and 10 have period 6, 12 has period 2. All are such that $c^{12} \equiv 1 \bmod 13$.

We saw that multiplication of R^+ by any of its elements yields a permutation of R^+. But consider now a prime p with a non-primitive root c with cycle m, which has to be a submultiple of p-1. Let the residues generated by c^n be $S=(s_1,s_2,\ldots,s_m)$, where $s_1=c$ and $s_m \equiv 1$. To multiply S by any of its constituents is to permute S and, thus, to do this systematically starting with multiplier s_1 and working through to s_m, yields each member of S exactly m times. If next we choose another residue d not contained in S and multiply each member of S by it we shall obtain a new set $T=(t_1,t_2,\ldots,t_m)$, say, in which all t_i are different from all s_j. Again, multiplication of T by each of t_i in turn produces a permutation of T, the result of which is that each member of T appears m times. This procedure may be continued until none of the original set R^+ remains, and the result is that each member of R^+ has appeared exactly m times, no more and no less.

References

1. Vajda, S. (1989) *Fibonacci and Lucas numbers, and the Golden Section.* Ellis Horwood. Chichester.

2. ***The Fibonacci Quarterly*** is devoted to our topic, and can be recommended to anyone who finds it as captivating as we do. In particular, fantasies of the kind described here, are likely to abound.

3. Hofstadter, D. (1979) *Gödel, Escher, Bach: an eternal gold braid.* The Harvester Press.

4. Mallows, Colin L. (1991) Conway's Challenge Sequence. Am. Math. Monthly, **98**,5-20.

5. Guy, Richard (1986) Some suspiciously simple sequences. Am. Math. Monthly, **93**, 186-190.

Table 1.2.2 The first 64 values of $H(n), F(n), C(n), N(n)$

n	$H(n)$	$F(n)$	$C(n)$	$N(n)$	n	$H(n)$	$F(n)$	$C(n)$	$N(n)$
1	1	1	1	1	32	17	16	16	16
2	1	1	1	1	33	17	17	17	16
3	2	2	2	2	34	20	18	18	16
4	3	2	2	2	35	21	18	19	16
5	3	3	3	2	36	19	19	20	16
6	4	4	4	3	37	20	20	21	17
7	5	4	4	4	38	22	20	21	18
8	5	4	4	4	39	21	20	22	18
9	6	5	5	4	40	22	21	23	19
10	6	6	6	4	41	23	22	24	20
11	6	6	7	5	42	23	22	24	20
12	8	7	7	6	43	24	23	25	20
13	8	8	8	6	44	24	24	26	21
14	8	8	8	7	45	24	24	26	22
15	10	8	8	8	46	24	24	27	22
16	9	8	8	8	47	24	24	27	23
17	10	9	9	8	48	32	25	27	24
18	11	10	10	8	49	24	26	28	24
19	11	10	11	8	50	25	26	29	24
20	12	11	12	9	51	30	27	29	24
21	12	12	12	10	52	28	28	30	25
22	12	12	13	10	53	26	28	30	26
23	12	12	14	11	54	30	28	30	26
24	16	13	14	12	55	30	29	31	27
25	14	14	15	12	56	28	30	31	28
26	14	14	15	12	57	32	30	31	28
27	16	15	15	13	58	30	31	31	28
28	16	16	16	14	59	32	32	32	29
29	16	16	16	14	60	32	32	32	30
30	16	16	16	15	61	32	32	32	30
31	20	16	16	16	62	32	32	32	31
					63	40	32	32	32
					64	33	32	32	32

Table 1.2.3 Values of H(n) at Powers of 2

n	$H(n)$
2	1
4	3
8	5
16	9
32	17
64	33
128	64
256	123
512	256
1024	557
2048	1047
4096	2066

1.3 FACING THE CLOCK

1.3.1 The clock on our church tower has an hour hand (H) and a minutes hand (M). We are interested in the configuration of these hands. To identify a point on the circumference of the clock face count the number of minutes positively clockwise, starting from 12 o'clock; the length of the circumference is thus 60, corresponding to 360°. At time x hours and y minutes, which we may denote by (x,y), where x is an integer such that $0 \leq x \leq 11$ and y is a real such that $0 \leq y < 60$, M is at y and H is at $5x+y/12$, since M moves twelve times as fast as H does. To every genuine time (x,y) the angle between H and M measures $5x$-$11y/12$ minutes, but if we were to move H and M to arbitrary positions they might not point to a genuine time. If M is at 0 (full hour), H must be at a multiple of 5. If the angle measures α and x is given, then $y=(60x-12\alpha)/11$. In practice the value of y used should be the least residue modulo 60. We have for each α a total of 11 different positions corresponding to $x=0,1,\ldots,10$. When M and H overlap, $\alpha=0$ and both hands have coordinate $60x/11$. Now let $\alpha=5t$, where t is a positive integer. This is the case at every full hour. Then

$$y=60(x-t)/11 \; = \; 60t'/11,$$

say, where t' is also an integer. This means that when $\alpha=5t$, the possible positions of M are precisely the 11 positions where M and H overlap.

We ask now whether any configuration is reversible, that is if H and M are given, corresponding to a time (x,y), does replacing H by M and M by H represent a true time? Clearly not all configurations are reversible. Think, for instance, of a full hour, not twelve o'clock.

Let M be at y and H at $5x+y/12$. Let the reverse configuration correspond to time (x',y'). Then $y'=5x+y/12$ and $5x'+y'/12=y$. Solving these for y and y' in terms of x and x'' gives

$$y=\frac{60}{143}(x+12x'), \quad y'=\frac{60}{143}(12x+x'). \tag{1.3.1}$$

For checking purposes it is useful to know that

$$y+y'=\frac{60}{11}(x'+x), \quad y-y'=\frac{60}{13}(x'-x). \tag{1.3.2}$$

The formulae also show that the times of reversible configurations are 60/143 minutes apart.

Examples (i) $x=5$, $x'=3$.

The formulae give $y=17\ 29/143$, $y'=26\ 62/143$. The diagram illustrates the two configurations.

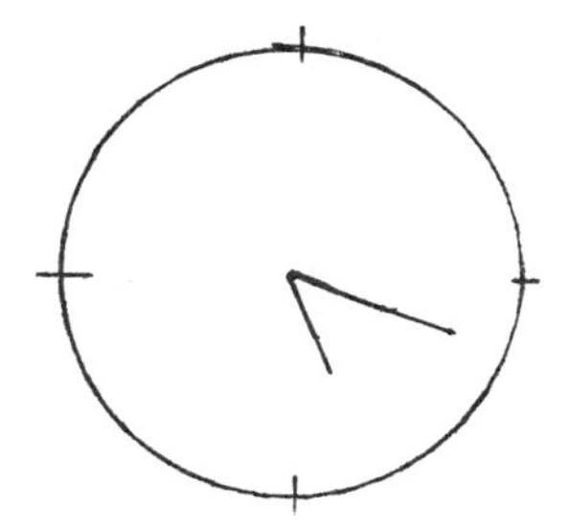

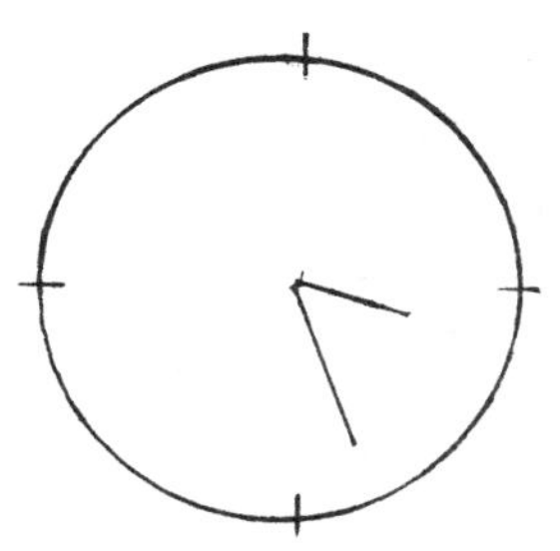

(ii) $x=11$, $x'=4$.

The formulae give $y=24\ 108/143$, $y'=57\ 9/143$, illustrated by the diagram below.

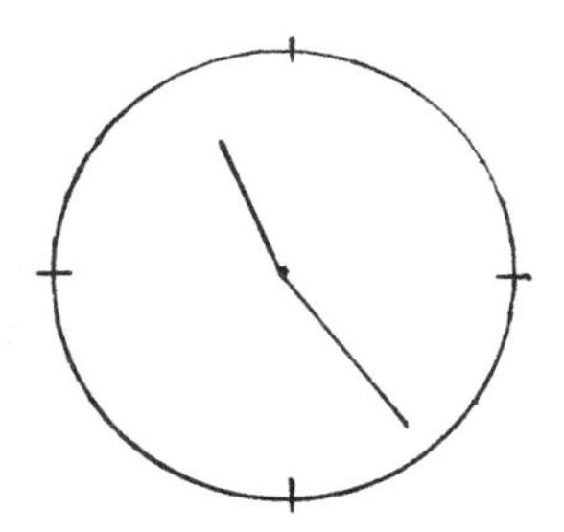

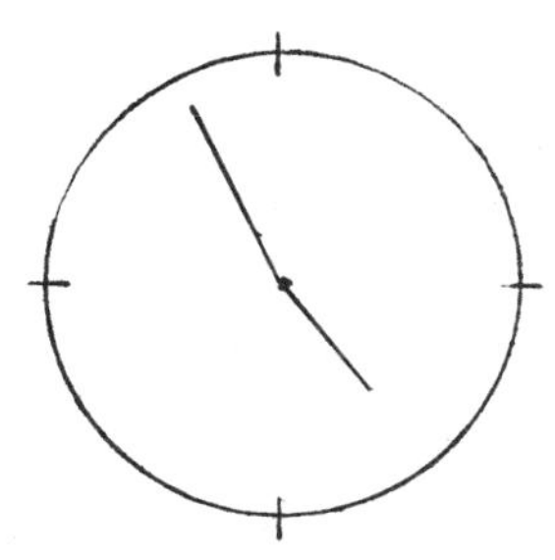

1.3.2 I awake in the night and snatch at my wristwatch, removed because of the heat. It has a luminous dial, but I am unable to distinguish between hour and minute hands, and I can not see the seconds hand at all. After squinting for some moments I decide that the time is about seven minutes to twelve. But it can't be, for I did not go to bed until half past twelve. I turn on the light and realise that I have been looking at my watch upside down. The real time is about half past four.

If the real time is (x,y), the coordinates (H,M) of the hour and minute hands are $H=5x+y/12$, $M=y$. Inversion of the clockface shifts the origin from 0 to 30. The coordinates (H',M') with respect to this origin are $H'=H-30$, $M'=M-30$, and the corresponding apparent time (x',y') is given by $H'=5x'+y'/12$, $M'=y'$. Now if $H'=H-30$, $M'=M-30$,

$$5x'+ y'/12=5x + y/12 - 30, \; y' = y-30,$$

giving

$$x'=x-11/2, \; y'=y-30.$$

This gives no real time since x' is not an integer when x is an integer, as it would be. But what if hour and minute hands are mistaken for each other in the dark? Then

$$5x'+y'/12=y-30, \; y'=5x+y/12 -30,$$

giving

$$y=\{720x'+60x+3960\}/143, \; y'=\{720x+60x'-3960\}/143.$$

For every pair of integers x, x' correspond a pair y and y', the sign of the latter having to be taken into account. For example, take $x=3$ and $x'=10$. These give $y=79.30$, $y'=-8.39$, and the corresponding times are (3,79.3) or (4,19.3), and (10,-8.4) or (9,51.6). Another, rather obvious, example is 6 o'clock. A search for the corresponding x and x' discloses $x=5$ and $x'=6$, when $y'=0$ and $y=60$. This freak result suggests searching for similar cases, but an investigation of the Diophantine equations for x and x' shows that

$y'=0$ only when $x=5$, $x'=6$;
$y=60$ only when $x=5$, $x'=6$;
$y'=60$ is not possible;
$y=0$ is not possible.

The clock in my bedroom hangs above my bed. On the opposite wall is a mirror in which I can see an image of the clock face. The time depicted by the mirror image looks real enough and the reader should find it easy to verify that it does indeed represent a genuine possible time. Indeed, if the real time is (x,y), the mirror tells me that it is $(11-x,60-y)$. No wonder that Alice had such a good time through the looking glass.

1.3.3 Impressed by the fact that a simple timepiece can stimulate so much arithmetic, I am again looking at my watch and I notice that it has a third hand which, perhaps confusingly, is called the seconds hand (S). Time is now given as x hours, y minutes and z seconds, (x,y,z), where x is, as before a positive integer with range $0 \leq x < 12$, but now y is also an integer in the range $0 \leq y < 60$, while z is real and $0 \leq z < 60$. The H, M and

S coordinates of the hours, minutes and seconds hands are given by

$$H=5x+\frac{y}{12}+\frac{z}{720},\quad M=y+\frac{z}{60},\quad S=z. \qquad (1.3.3)$$

For the three hands to coincide, H=M=S, and this implies that

$$z=\frac{60x}{11},\quad y=\frac{59x}{660}.$$

Apart from $x=0$ there is no admissible value of x that makes y an integer and it follows that **the only time when the three hands coincide is 12 o'clock.** It is nevertheless possible to find times when two out of H,M,S are reversed. We saw how to find configurations (H,M) and (M,H), but each of these implies a different S. From equations (1.3.1) we see that the seconds value z is given by

$$z=60(y-[y])$$

where $[y]$ is the integer part of y. The value z' is obtained similarly.

Again, given y and y', we can find z and z' corresponding to the configurations (H,M,S) and (H',S,M). For now

$$M': y'+\frac{z'}{60}=z,\qquad S': z'=y+\frac{z}{60},$$

leading to

$$z=\frac{3600}{3599}(y'+\frac{y}{60}),\qquad z'=\frac{3600}{3599}(y+\frac{y'}{60}).$$

An arbitrary assignment of integer minute values y,y' leads to real values z and z' and these may be associated with arbitrary x and x' to give the coordinates of H and H'. For example, with $x=1,x'=11,y=10,y'=35$ we get the (M,S)-reversible times (1,10,35.1764),(11,35,10.5863).

Similarly, for the reversible (H,M,S)/(S,M',H) we need

$$5x' + \frac{y'}{12} + \frac{z'}{720} = z, \quad 5x + \frac{y}{12} + \frac{z}{720} = z',$$

giving

$$z = \frac{1}{518399}[3600(720x' + x) + 60(720y' + y)],$$

with a similar expresssion for z'. An example is $x=1, y=10, x'=11, y'=35$ giving $z=57.9249$, $z'=5.9138$. However, since in this case x and y are constrained to be integers, apart from the time (0,0,0), there can be no reversible configurations in which the positions of all three hands are permuted.

1.4 ENCOUNTERS

1.4.1 A well-known sociologist is studying a community of men and women. He wishes to test a theory concerning the chance meetings of opposite sexes at a rate proportional to the product of the numbers of each sex available in the community. Such meetings are also called **encounters** and the sociologist is particularly interested in three distinct types:

(i) those entailing no commitment;a simple friendship, for example: this is in contrast with

(ii) those in which the couple engage to marry and are therefore removed from the eligible pool of sexes. These are called **monogamous** encounters;

(iii) **polygamous** encounters; in this case, one or other of the sexes is allowed unlimited partners and the eligible pool is reduced by one each time a 'marriage' takes place.

If a meeting or encounter is also described as an **event**, a word probabilists are fond of, the investigation requires knowledge of the statistics of

(a) the time S_n to the n-th event, and

(b) the number N(t) of events that have occurred in given time interval $(0,t)$.

To develop a theory our sociologist is advised by statistical friends to lay a foundation as follows: that when n enounters have taken place (n an integer not less than 0), and measuring time from an arbitrary starting point, the probability that the $(n+1)$-th encounter occurs in the small time interval $(t, t+h)$ later, is $a_{n+1} h + o(h)$, where a_{n+1} is independent of t but dependent on $n+1$, and $o(h)$ is shorthand for terms of the order $h^{1+\alpha}$, $\alpha>0$. This means that the future evolution of the encounter process depends only on the number of encounters that have so far occurred. To adapt to the types (i), (ii), (iii) of encounter, the constants a_n will be made to depend in an appropriate way on the available number of the sexes eligible.

Many, if not most, readers will recognise that the comprehensive model implies an exponential distribution of the intervals T_n between the *(n-1)-th* -and *n-th* encounters $(n \geq 1)$. This is explained in the Notes. When all a_n are constant, the study of S_n and $N(t)$

is that of a Poisson process with constant parameter. Note that in general a_n can be interpreted as a mean rate at which events subject to the conditions implied by the form of a_n take place.

Let S_n have probability density function $b_n(t)$, and let $N(t)$ have point probability $f_n(t)$. Then , by definition,

$$b_n(t)dt = dP(S_n<t),$$

$$f_n(t) = P[N(t)=n|N(0)=0] \qquad (1.4.1)$$

The Notes below show that

$$b_n(t)=a_1a_2...a_n\sum_{m=1}^{n} c_m e^{-a_m t} \quad (n\geq 1) \qquad (1.4.2)$$

where

$$c_m^{-1}=(a_{m+1}-a_m)(a_{m+2}-a_m)...(a_n-a_m)(a_1-a_m)(a_2-a_m)...(a_{m-1}-a_m) \quad (m\geq 1);$$

and that

$$f_n(t)=P[S_n<t,T_{n+1}>t-S_n]$$

$$=\int_0^t b_n(s)e^{-a_{n+1}(t-s)}ds. \qquad (1.4.3)$$

The form of a_n arises from the application and usually imposes a restriction on n and m.

Armed with these results the sociologist turns to his models (i)...(iii).

1.4.2 Non-committal encounters

In this case an encounter does not disqualify the participants from further encounters, with each other or with other members not previously encountered. The average encounter rate is taken as proportional to the product of the numbers of men (M) and women (W) available. To fix the time scale a constant p is introduced, interpreted as the mean rate at which a population of one male and one female meet each other. If this is once per time unit, $MWpt$ is the mean number of encounters in time $(0,t)$. Thus $a_n=MWp=a$, independent of n. The encounter model is a straight Poisson process with mean rate a, and therefore

$$b_n(t) = ae^{-at}\frac{(at)^{n-1}}{(n-1)!},$$

$$f_n(t) = e^{-at}\frac{(at)^n}{n!}, \quad (n \geq 0). \qquad (1.4.4)$$

In this case n is unlimited.

1.4.3 Monogamous encounters

Now an encounter results in the removal of one individual of each sex. If the initial size is M men and W women, these are reduced to M-1 and W-1 by the first encounter. Thus $a_1 = MWp$, and $a_2 = (M-1)(W-1)p$. Successive encounters reduce the population size until the N-th encounter, where $N = \min(M, W)$, and no more marriages are possible. The formulae, restricted for tidiness to the case $M = W = N$, are based on

$$a_n = (N-n+1)^2 p$$

for $1 \leq n \leq N$, since the encounter process terminates at the N-th encounter, in this case marriage. This mechanism leads via (1.4.2) and (1.4.3) to:

$$b_n(t) = n^2 p \binom{N}{n}^2 \sum_{i=0}^{n-1} \frac{(-1)^{n-i-1}\binom{n-1}{i}^2 e^{-(N-i)^2 pt}}{\binom{2N-2i-1}{n-i-1}\binom{2N-i}{i}}, \qquad (1.4.5)$$

and

$$f_n(t) = \binom{N}{n}^2 \sum_{i=0}^{n} \binom{n}{i}^2 \frac{(-1)^{n-i} e^{-(N-i)^2 pt}}{\binom{2N-i}{i}\binom{2N-2i-1}{n-i}}, \quad (0 \leq n \leq N). \qquad (1.4.6)$$

The probability that the population is completely paired off by time t is

$$f_N(t) = 1 - 2\sum_{n=1}^{N} (-1)^{n-1} \frac{\binom{2N}{N-n}}{\binom{2N}{N}} e^{-n^2 pt}. \qquad (1.4.7)$$

The mean and second moment of S_N have interesting values as $N \to \infty$.

1.4.4 Polygamous encounters

Here it is supposed that a man can take multiple wives. When a marriage occurs the female population is decreased by 1, but the male population is fixed, at M, say. We shall let the initial number of women be N. Then $a_1=MNp$ and, writing $Mp=q$, we have

$$a_n=(N-n+1)q, \quad (1\leq n\leq N).$$

The process terminates when n reaches N. From the general formulae we get

$$b_n(t)=Nq\binom{N-1}{n-1}e^{-Nqt}\sum_{m=1}^{n}(-1)^{n-m}\binom{n-1}{m-1}e^{(m-1)qt}$$

$$=Nq\binom{N-1}{n-1}e^{-(N-n+1)qt}(1-e^{-qt})^{n-1}, \qquad (1.4.8)$$

with

$$f_n(t)=\binom{N}{n}e^{-(N-n)qt}(1-e^{-qt})^{n}. \qquad (1.4.9)$$

The sociologist asks himself whether the binomial form of the probabilities could have been foreseen. Could it?

The same mechanism applies to a population infiltrated by the carrier of an infectious disease. If the population consists initially of N susceptible individuals and each infected individual is isolated as soon as the disease shows, but the carrier, a polluted stream for instance, is not identified, the susceptible population decreases from N by steps of 1, and $a_1=Np$, $a_2=(N-1)p,\ldots,a_n=(N-n+1)p,\ldots$, $a_N=p$, whereupon there are no susceptibles left. The practical importance to society of this application is perhaps greater than that of polygamous habits.

1.4.5. Epidemic without Isolation

The disease-free population in this example is infiltrated by an infected individual who passes the illness on to others. These, now carriers and not being isolated, perhaps because they can not be identified, in turn infect the remainder of the susceptible population. It is convenient to take $2N$ as the initial size of the susceptible population. The mechanism is: $a_1=1.2Np, a_2=2.(2N-1)p,\ldots,a_n=n(2N-n+1),\ldots,a_N=N(N+1)$, but then $a_{N+1}=a_N$, $a_{N+2}=a_{N-1},\ldots,a_{2N}=a_1$. The total time S_{2N} for the disease to sweep through the population is thus the sum of two random variables each being distributed as S_N. The basic results are:

$$b_n(t)=2Np\sum_{m=1}^{n}(-1)^{m-1}(2N-2m+1)\frac{\binom{n}{m}\binom{2N-1}{m-1}}{\binom{2N-n}{m}}e^{-m(2N-m+1)pt}, \qquad (1.4.10)$$

$$f_n(t)=2Npe^{-(n+1)(2N-n)pt}\sum_{m=1}^{n}\frac{(-1)^{m-1}(2N-2m+1)\binom{n}{m}\binom{2N-1}{m-1}}{(n-m+1)(2N-n-m)\binom{2N-n}{m}}x$$

$$x[e^{(n-m+1)(2N-n-m)pt}-1],\ (1\leq n\leq N). \qquad (1.4.11)$$

The duration S_{2N} of the epidemic has probability density function $b_{2N}(t)$ obtained by convolving $b_N(t)$ with itself.

1.4.6 Notes

Let $E_1,E_2,\ldots,E_n,\ldots$ be a sequence of events evolving in time t. Starting from an arbitrary time origin, let T_1 be the time to E_1. Let $P(t<T_1<t+h)=a_1h+o(h)$, where a_1 does not depend on t or h. $o(h)$ is shorthand for a term such that $\lim_{h->0} o(h)/h = 0$. Let $f_0(t)=P(T_1>t)$, that is, the probability that E_1 does not happen until after time t has passed. Then

$$f_0(t+h) = f_0(t)(1\text{-}a_1h)+o(h)$$

since $P(T_1>t+h)$ is, by assumption, the product of the probabilities of the independent events $T_1>t$ and 'E_1 does not occur in $(t,t+h)$'. Expansion of the left hand side near t leads to $hdf_0(t)/dt=\text{-}a_1hf_0(t)+o(h)$, and then division by h followed by $h\text{-}>0$, gives the differential equation

$$\frac{df_0}{dt}=-a_1f_0(t), \qquad (1.4.12)$$

with solution

$$f_0(t)=e^{-a_1t} \qquad (1.4.13)$$

because $f_0(0)=1$. Since $P(t<T_1<t+h)=f_0(t)$ multiplied by $P(E_1\ )$ occurs in $(t,t+h)$ it follows that the probability density function $b_1(t)$ of T_1 is given by

$$b_1(t)=f_0(t)a_1=a_1e^{-a_1t}. \qquad (1.4.14)$$

Thus T_1 is a continuous, non-negative exponentially distributed random variable, **and it does not matter from which time origin we start to measure it provided that the conditions governing the parameter a_1 do not change**. Next, starting at the occurrence of E_1, the argument can be repeated to find the probability density function of the time interval T_2 between E_1 and E_2, with constant a_2 replacing a_1. And so on for T_3, T_4,... Each is exponentially distributed.

The text requires general formulae for probabilities associated with the sum

$$S_n = T_1 + T_2 + \ldots + T_n,$$

and with the number $N(t)$ of events that have occurred in $(0,t)$. These are related. Let $b_n(t)$ be the probability density function of S_n, that is,

$$b_n(t)dt = dP(S_n<T), \qquad (1.4.15)$$

and

$$f_n(t) = P(S_n<t, T_{n+1}>t-S_n). \qquad (1.4.16)$$

Since S_n and T_{n+1} are independent, and

$$S_{n+1} = S_n + T_{n+1},$$

$$b_{n+1}(t)=a_{n+1}\int_0^t b_n(u)e^{-a_{n+1}(t-u)}du, \quad (n\geq 1), \qquad (1.4.17)$$

and

$$f_n(t)=\int_0^t b_n(u)e^{-a_{n+1}(t-u)}du, \quad (n\geq 1). \qquad (1.4.18)$$

In particular, if $n=1$, since $b_1(u)=a_1\exp(-a_1u)$,

$$b_2(t) = a_1 a_2 \int_0^t e^{-a_1 u} e^{-a_2(t-u)} du$$

$$= \frac{a_1 a_2}{a_2 - a_1}\left(e^{-a_1 t} - e^{-a_2 t}\right),$$

and

$$f_1(t) = a_1 \int_0^t e^{-a_1 u} e^{-a_2(t-u)} du$$

$$= \frac{a_1}{a_2 - a_1}\left(e^{-a_1 t} - e^{-a_2 t}\right). \quad (1.4.19)$$

Since the Laplace transform of convolutions such as (1.4.17) and (1.4.18) is the product of the Laplace transforms of the functions under the integral signs, we get

$$\beta_n(z) = \frac{a_1 a_2 ... a_n}{(z+a_1)(z+a_2)...(z+a_n)}, \quad (1.4.20)$$

where

$$\beta_n(z) = \int_0^\infty e^{-zt} b_n(t) dt$$

is the Laplace transform of $b_n(t)$. Decomposition of (1.4.20) into the component partial fractions and inversion back to the time domain gives the key formula

$$b_n(t) = a_1 a_2 ... a_n \sum_{m=1}^{n} c_m e^{-a_m t}$$

where

$$c_m^{-1} = (a_{m+1} - a_m)(a_{m+2} - a_m)...(a_n - a_m)(a_1 - a_m)(a_2 - a_m)...(a_{m-1} - a_m). \quad (1.4.21)$$

These are the general formulae used to give the particular results in the text.

1.5 THE HYPERGEOMETRIC DISTRIBUTION AND SOME APPLICATIONS

1.5.1 Consider a population of N members belonging to one or other of two types, A and B, say. Let there be a A-types and therefore $b=N-a$ B-types. A sample of size n is drawn "randomly" (meaning that all possible samples have the same probability) without replacement. I want the probability $H(N,n,x)$ that the sample contains x A-types. One of the reasons for interest in this distribution is its application to **Quality Control.** Here a batch of N items is manufactured, some of which may be defective. If too many are defective the manufacturer desires to jettison the whole batch, but there are often too many items to test them individually, or, testing may entail destroying an item; in either case it is convenient to draw a sample from the batch, to test the items in the sample, and then to draw inferences about the number of defectives in the batch. **Opinion Polls** provide another example. It is not usually practical to ask thousands of people whom they intend to vote for, or which soap they prefer, so the pollsters again make inferences from a sample. Here we shall refer also to questions arising in **The Lottery** as introduced in Great Britain in 1994.

1.5.2 The probability $H(N,n,x)$ referred to above is given by

$$H(N,n,x)=\frac{\binom{a}{x}\binom{N-a}{n-x}}{\binom{N}{n}}. \qquad (1.5.1)$$

For convenience in the text we denote the

$$\textit{binomial coefficient} \binom{N}{n} \textit{ by } [N,n].$$

To explain the form of H it is merely necessary to observe that there are $[N,n]$ combinations of n that can be selected from N items: random selection means asigning each the same weight. The number of combinations of x out of a A-types is $[a,x]$ and each of these must be multiplied by the number of combinations of n-x drawn from N-a B-types. Thus (1.5.1). The mean and variance of the random variable X $(=x)$ are given by

$$E(X)=\frac{an}{N}, \qquad Var(X)=\frac{(N-n)(N-a)an}{N^2(N-1)}.$$

Note that if as $N\to\infty$, $n\to pN$ $(0<p<1)$, then

$$E(X)=ap, \qquad Var(X)=ap(1-p),$$

and indeed under these circumstances the hypergeometric distribution tends to the binomial

$$\binom{a}{x}p^x(1-p)^{a-x}.$$

The reason for the name "hypergeometric" is the connection betweeen H and terms in the series representation of the Gauss hypergeometric function

$$F(a,b;c;y)=1+\frac{ab}{c}\frac{y}{1!}+\frac{a(a+1)b(b+1)}{c(c+1)}\frac{y^2}{2!}+\ldots$$

and in fact $H(N,n,x)$ is generated by

$$\frac{\binom{N-a}{n}}{\binom{N}{n}}F(-n,-a;N-a;x).$$

1.5.3 Quality control The following is the theory behind a particular quality control procedure which requires careful logical formulation. A manufacturer produces ball-bearings in batches of N items. The batch is tested in the following way. Single items are examined and classified as either effective or defective. Testing stops at the latest when $n(\leq N)$ items have been picked at random, or earlier if either $c+1$ $(\leq n)$ defective items have been found, when the batch is rejected: or if it becomes impossible to find $c+1$ defective items, even if testing were continued, when the lot is accepted. We want to know the expected number of items which will be tested if there are m defective items in the batch. m may be $>$ or $\leq c$.

Sampling terminates by rejection after $c+i+1$ steps $(i=0,1,\ldots,n-c-1)$, if c defectives have been found among the first $c+i$ items, and one more item is found to be defective. The probability of this is

$$\frac{\binom{m}{c}\binom{N-m}{i}}{\binom{N}{c+i}}\frac{m-c}{N-c-i}=\frac{\binom{c+i}{c}\binom{N-c-i-1}{m-c-1}}{\binom{N}{m}} \qquad (1.5.3.1).$$

This is zero when $m \leq c$, as it obviously must be.

Sampling terminates with acceptance after $n\text{-}c+j$ $(j=0,1,\ldots,c)$ steps if there are j defectives among the first $n\text{-}c+j\text{-}1$ items, and the last of these is effective, because then only $c\text{-}j$ items are left to be looked at, and they can not contain $c\text{-}j+1$ more defectives, which would be necessary for rejection. The probability of this is

$$\frac{\binom{N-m}{n-c-1}\binom{m}{j}}{\binom{N}{n-c+j-1}}\frac{N-m-n+c+1}{N-m+c-j+1}=\frac{\binom{N-n+c-j}{m-j}\binom{n-c+j-1}{j}}{\binom{N}{m}}. \qquad (1.5.3.2)$$

This is zero when $m \geq N\text{-}n+c+1$.

The average sampling number S for given N,n,m and c is given by

$$\binom{N}{m}S=\sum_{i=0}^{n-c-1}(c+i+1)\binom{c+i}{c}\binom{N-c-i-1}{m-c-1}+\sum_{j=0}^{c}(n-c+j)\binom{n-c+j-1}{j}\binom{N-n+c-j}{m-j}.$$

This can also be expressed as

$$\binom{N}{m}S=(c+1)\sum_{i=0}^{n-c-1}\binom{c+i+1}{c+1}\binom{N-c-i-1}{m-c-1}+(n-c)\sum_{i=0}^{c}\binom{n-i}{n-c-1}\binom{N-n+i}{m-c+i}.$$

If m=0, S=n-c because only the term for i=c remains in the second sum; if m=N, only the term for i=0 remains in the first sum and S=c+1. This is a check verified by the meaning of the symbols. Some numerical values are given in [1].

1.5.4 Lottery lore

The National Lottery introduced into Great Britain in 1994 is operated roughly as follows. Each week, six integers from the set 1..49 are selected at random without replacement. Those gamblers who have forecast at least three correctly are eligible for a prize. The more correct, the greater the potential prize. From what has been written in 1.5.1 the reader will recognize that N=49 and $n=a$=6. The probability of getting x right is $H(N,n,x)$, and the values are:

x	0	1	2	3	4	5	6
H	0.4360	0.4130	0.1324	0.0176	0.0010	$2x10^{-5}$	$72x10^{-9}$

Not very encouraging for those hoping for six. Evidently the statisticians have been at work! We do not offer in this essay a recipe to ensure a win (which would be impossible), but we look at a fragment of the many intriguing problems that might be attacked. How many weeks, for instance, does it take for all the set 1..49 to appear? Let M_t be the number of different integers produced by the t-th lottery draw, and let

$$p_m^{(t)}=P(M_t=m).$$

Then $p^{(1)}{}_n=1$ and $p^{(2)}{}_m$ has nonzero values for $m=n,n+1,\ldots,2n$. These probabilities are generated by the powers of x in

$$G_t(x)=\sum_{r\geq 0} P(M_t=n+r)x^r,$$

and it can be shown by standard methods that $H_t(x)=x^nG_t(x)$ satisfies the recursive integral equation

$$\binom{N}{n}H_{t+1}(x)=\frac{1}{2\pi i}\oint(1+xz)^N H_t\{\frac{x(1+z)}{1+xz}\}\frac{dz}{z^{n+1}}.$$

the contour being the unit circle. As a check, note that putting $x=1$ and using $H_t(1)=1$ (as it should) implies that $H_{t+1}(1)=1$. From the generating function can be obtained the expected value and variance of M_t, by differentiating the integral equation and solving the resulting elementary difference equation. The results are:

$$E(M_t)=N\ \{1-(1-\frac{n}{N})^t\},$$

$$Var(M_t)=N(1-\frac{n}{N})^t\ \{1-N(1-\frac{n}{N})^t+(N-1)(1-\frac{n}{N-1})^t\ \}.$$

We see that as $t\text{-}>\infty$, $M_t\text{-}>N$, but that it takes quite a long time.

If the lottery randomizing mechanism is *really* random it should, over a long period, generate each of 1..49 about the same number of times. Gamblers might therefore believe it worth while to bet on numbers with a deficiency of appearances. The probability of any of the N possible digits turning up in a draw is n/N. For, consider the digit 1: at any draw the probability that 1 is the first number generated is $1/N$, and the probability that it is the second is $(1-1/N)/(N-1)=1/N$, so the probability that it is either the first or the second is $2/N$. Continuing shows that the probability that 1 appears in a draw is n/N. This holds for all possible 49 digits. Thus, the probability that in t lottery draws a given number appears r times is $[t,r](n/N)^r\ (1-n/N)^{t-r}$, and its expectation is tn/N and variance $tn(1-n/N)/N$. At the moment of writing there have been 18 draws. The frequency of the 49 numbers (excluding the so-called "bonus number") is:

#	1	2	3	4	5	6	7	8	9	10
frq	1	3	3	1	4	2	2	1	3	1
#	11	12	13	14	15	16	17	18	19	20
frq	2	2	2	2	2	4	4	3	2	1
#	21	22	23	24	25	26	27	28	29	30
frq	3	3	1	1	2	3	2	0	5	3
#	31	32	33	34	35	36	37	38	39	40
frq	5	3	1	0	2	3	1	5	1	1
#	41	42	43	44	45	46	47	48	49	
frq	1	4	2	4	1	2	2	1	1	

The expected frequency is 2.2, and standard deviation 1.4. There is nothing remarkable about the results. An optimist may like to bet on numbers with below-average frequency, but do not hold your hopes excessively high.

Reference

1. Vajda,S. (1946) Average sampling numbers from finite lots. Suppl. J.Roy. Statist. Soc., **8**, 198-201.

2

Financial

This chapter deals with two aspects of financial mathematics: the first describes a scheme introduced after the first world war in Germany, Austria and Switzerland to enable savers to finance house purchase; the second is concerned with reinsurance, a procedure whereby insurance companies try to protect themselves against large, unforeseen claims; this study involves statistical estimation procedures. Such applications of mathematics are, perhaps, not as widely known as they should be and should therefore stimulate interest, especially among graduands looking for a career.

2.1 COLLECTIVE DIY FINANCING

2.1.1 Imagine a Building Society which accepts contributions only from members who intend eventually to receive loans of fixed constant amount £H for house purchase from the Society's funds. Since H is to be constant we shall replace it by unity. A member's contributions are credited to his account during a *'waiting time'* of duration w, which varies fron member to member. The amount a member borrows can exceed his current balance, and he repays the loan, i.e. the difference between the amount standing to his credit and the amount he needs, during a *'repayment time'"* of duration r. The repayments are credited at the same interest rate as the deposits, during the waiting time. The total $m=w+r$ is the *membership time.*

The deposits and repayments may be made annually, monthly, or at other, possibly irregular intervals. We shall assume that regular payments of amount $p\Delta t$ are made at time intervals Δt, and we shall approximate the precise development of credits and debits by letting Δt tend to zero. This is a convenient way of accounting and gives a clear picture of the ups and downs in the life of the Society. In the actuarial world p is called the *force, or intensity of payment,* and we shall make use of this description.

Let the intensity of payment by a member during w be α, and during r let it be β. Then, if $c=\ln(1+i)$, i being the nominal annual rate of interest, the member will, when he receives his loan at the end of w, have accumulated a credit balance of

$$\frac{\alpha}{c}(e^{wc}-1)=A, \qquad (2.1.1)$$

as shown in the Notes at the end. The loan required is $B=1$-A. To repay B, the member pays during r, $\beta\Delta t$ at regular intervals Δt. It is shown in the Notes that if Δt->0, β must be such that

$$\frac{\beta}{c}(1-e^{-rc})=B. \qquad (2.1.2)$$

Since $A+B=1$,

$$k = \beta - \alpha - c = \beta e^{-cr} - \alpha e^{cw}.$$

Hence

$$e^{-cr} = \frac{(\alpha e^{cw} + k)}{\beta}, \tag{2.1.3}$$

$$r = \frac{1}{c} \ln \frac{\beta}{\alpha e^{cw} + k}, \tag{2.1.4}$$

and

$$m = r + w = \frac{1}{c} \ln \frac{\beta}{\alpha e^{cw} + k} + \frac{1}{c} \ln e^{cw} = \frac{1}{c} \ln \frac{\beta}{\alpha + k e^{-wc}}. \tag{2.1.5}$$

In the above, k is a constant. If $w=0$, meaning that the member borrows the full amount without any preliminary saving, then

$$m = r = \frac{1}{c} \ln \frac{\beta}{\beta - c}.$$

The case $k=0$ is said to determine the *harmonic tariff* and merits special interest. In this case equation (2.1.5) reduces to

$$m = \frac{1}{c} \ln \frac{\beta}{\alpha} = \frac{1}{c} \ln \frac{\alpha + c}{\alpha},$$

which is independent of w. This independence can be understood by arguing as follows: if a member pays α all the time, it takes

$$\frac{1}{c} \ln \left[\frac{c}{\alpha} + 1 \right]$$

time units to accumulate the unit house-purchase price. This follows from (2.1.1), setting $w=m$. If a member receives the loan after $w<m$, we can imagine that he does not touch his accumulated A, but carries on paying α with c as interest, having received the unit loan too early. So he pays at rate $\alpha+c=\beta$, and when he has accumulated 1, after m time units, he pays 1 back. His total membership time lasts just as long as necessary to accumulate the unit amount by payments of α.

Formulae (2.1.1) to (2.1.5) refer to individual members; loans are, of course, possible only if the Society has several members. Assume, then, that there is a continuous flux of new members with constant intensity λ. Loans are allocated to members in order of entry time. For further analysis of the Society's membership we consider first the initial period ($0 \le y \le Y$) in the history of the Society, during which no member has yet completed his membership time.

During this period members are either saving or repaying their loans. Specifically Y is the time when the initial member finishes repaying his loan, and $Y=m$ is given by (2.1.5). Let a member receive his loan at time y, after having waited for a time w_y. This means that he joined the Society at time $y-w_y$. Since new members join at rate λ, λw_y is the number of members who entered after the instant when the member we are considering joined the Society, and since no-one has left they are all waiting. The number of waiters, that is savers, increases through new entries and decreases through loan allocations. Let the number of allocations during the short time interval $(y,y+h)$ be hz_y, so that z_y, a variable, is the intensity of loan allocations at time y. It follows that

$$\lambda w_{y+h}=\lambda w_y-hz_y+\lambda h.$$

When $h->0$ this reduces to the differential equation

$$\frac{dw_y}{dy}=1-\frac{z_y}{\lambda}. \qquad (2.1.6)$$

Since $dw_y/dy>0$ while $\lambda>z_y$, it follows that the waiting time increases when the intensity of entries exceeds the intensity of loan allocations. To fulfil its objectives, the Society must keep within its resources. This means that the rate of allocation must be balanced by the rate at which savers contribute plus the rate at which borrowers are repaying. The number of savers at time y is, as argued above, λw_y, and the number of members is λy, so that $\lambda(y-w_y)$ are repaying. Since with each member is associated a unit amount, and the saving and repayment rates are α and β, respectively, it follows that

$$z_y=\lambda\alpha w_y+\lambda\beta(y-w_y). \qquad (2.1.7)$$

Thus, from (2.1.6) and (2.1.7), we have

$$\frac{dw_y}{dy}+(\alpha-\beta)w_y=1-\beta y, \quad (0\leq y\leq Y). \qquad (2.1.8)$$

With $w_0=0$ we then get for $0\leq y\leq Y$,

$$w_y=y-\frac{\alpha}{(\beta-\alpha)^2}\left[e^{(\beta-\alpha)y}-(\beta-\alpha)y-1\right]. \qquad (2.1.9)$$

It follows from (2.1.6) that

$$z_y=\lambda(1-\frac{dw_y}{dy})=\frac{1}{\beta-\alpha}(e^{(\beta-\alpha)y}-1),$$

$$\textit{for } 0\leq y\leq Y. \qquad (2.1.10)$$

(2.1.9) and (2.1.10) describe the Society's membership during the initial period $0\leq y\leq Y$. We turn now to the period $y>Y$ when members are already leaving after termination of their membership time. Y is the time when the first member leaves and is given by (2.1.5). Denote by n_y the number of members, savers plus borrowers, at time $y(>Y)$. In this case (2.1.7) becomes

$$z_y=\alpha\lambda w_y+\beta(n_y-\lambda w_y), \qquad (2.1.11)$$

and

$$\frac{dw_y}{dy}+(\alpha-\beta)w_y=1-\frac{\beta n_y}{\lambda}. \qquad (2.1.12)$$

The solution of this differential equation is

$$w_y=\frac{1}{\alpha-\beta}+I\ e^{(\beta-\alpha)y}-\frac{\beta}{\lambda}e^{(\beta-\alpha)y}\int_Y^y e^{(\alpha-\beta)u}n_u\ du. \qquad (2.1.13)$$

The constant of integration I is determined from the value of w_y when $y=Y$. We obtain

$$w_Y=\frac{\beta Y}{\beta-\alpha}-\frac{\alpha}{(\beta-\alpha)^2}\{e^{(\beta-\alpha)Y}-1\}$$

and

$$I=(w_Y-\frac{1}{\alpha-\beta})e^{(\alpha-\beta)Y}. \qquad (2.1.14)$$

When $k=0$, the case of the *harmonic tariff*, we can give more detailed information. In this case we know already that

$$m=\frac{1}{c}\ln\frac{\beta}{\alpha}\quad (\textit{constant}).$$

Thus the first member to leave is the member who received his loan at time 0, and so

$$Y=\frac{1}{c}\ln\frac{\beta}{\alpha}. \qquad (2.1.15)$$

From (2.1.9) we get, using $k=\beta-\alpha-c=0$,

$$w_y=\frac{\beta y}{c}-\frac{\alpha}{c^2}(e^{cy}-1), \qquad (0\le y\le Y), \qquad (2.1.16)$$

and so, putting $y=Y$,

$$w_Y=\frac{\beta Y}{c}-\frac{\alpha}{c^2}\left(\frac{\beta}{\alpha}-1\right)=\frac{1}{c}(\beta Y-1). \qquad (2.1.17)$$

Since, when $k=0$, membership time $m=Y$ is constant for each member,n_u has the **constant** value λY, and evaluation of (2.1.14), using (2.1.17), gives for $Y>y$ the constant value

$$w_y=w_Y\ , \qquad (2.1.18)$$

and the number of repayers is

$$(n_y-\lambda w_y)=\lambda(Y-w_Y)=\frac{\lambda}{c}(1-\alpha\ln\frac{\beta}{\alpha}). \qquad (2.1.19)$$

Finally,for $y\geq Y$, z_y reduces to λ, confirming that a state of equilibrium has been achieved when the harmonic tariff is applied.

The scheme described is, in essence, that applied after World War 1 in Germany, Austria, and also in Switzerland, where money for new housing was, naturally, scarce. Since then the scheme has been changed for practical and for legal reasons, and consequently the interest in continuing the theoretical analysis waned. No complete description of developments when $k\neq 0$ is known to us. But it is an interesting fact that the first, and one of the largest of such companies, the Gesellschaft der Freunde, Wüstenrot, still operates from headquarters in Ludwigsburg near Stuttgart, with a namesake which is a bank, and another which is a life insurance company.

2.1.2 Notes

Let an asset n_t at time t yield interest in such a way that between t and $t+\Delta t$ the asset is increased to $n_t(1+c\Delta t)$. Then

$$\lim_{\Delta t=0} \frac{n_{t+\Delta t}-n_t}{n_t \Delta t} = \frac{dn_t}{n_t dt} = \frac{d\ln n_t}{dt} = c,$$

This implies that

$$n_t = n_0 e^{ct}.$$

The equivalent interest rate i for a unit period of one year is given by

$$e^c = 1+i, \; or \; c = \ln(1+i).$$

It follows that the total sum built up from payments of amount $\alpha\Delta t$ at a succession of intervals Δt over n years will be

$$\alpha \int_0^n e^{ct} dt = \frac{\alpha}{c}(e^{nc}-1),$$

and, similarly, the present, or discounted, value of payments $\beta\Delta t$ at intervals Δt over n years is

$$\frac{\beta}{c}(1-e^{-nc}).$$

2.2 REINSURANCE

Why do people insure their property against fire, accidents and other hazards? Because they prefer to pay a regular, fixed affordable premium rather than to suffer a possibly very heavy loss. Why can an insurer help? Because a large portfolio of assorted risks makes it easier to forecast the total of claims. However, even a large company will wish to limit its commitments, and therefore it *cedes* a portion of its portfolio to a *reinsurer*.

Many forms of reinsurance have been devised and practised. we shall deal here with a type of some mathematical interest: *Stop-loss reinsurance*. The reinsurer reimburses the excess of the total claim made against the *cedent* over an amount C, this excess being called the *priority*. Let the probability density function of the aggregate X of claims against the cedent during one year be denoted by $f(x,\tau)$, where τ is a parameter. X is by its nature a non-negative random variable which in this essay will be exponentially distributed with population mean τ, so that

$$f(x,\tau)=\frac{e^{-\frac{x}{\tau}}}{\tau}, \quad (0<x<\tau). \qquad (2.2.1)$$

In [1] X is treated as a discrete Poisson-distributed random variable.

The reinsurance premium paid for the contract will be based in some way, and this essay does not go into how, on the expected claim P given by

$$P=\frac{1}{\tau}\int_{C}^{\infty}(x-C)e^{-\frac{x}{\tau}}dx=\tau e^{-\frac{C}{\tau}}. \qquad (2.2.2)$$

But neither cedent nor reinsurance company has more to go on than past records and strong evidence, in this case, that $f(x,\tau)$ is exponential. Hence an estimate $\overline{P}$ of P must be made, based on, say, the record of claims $(x_1,\ldots,x_n)$ made in each of the n previous years. An obvious estimate is

$$\overline{P}=\frac{1}{m}\sum_{i\in I}(x_i-C), \qquad (2.2.3)$$

where I is the set of, say, m integers for which $x_i>C$. But since I may be empty this estimate could prescribe a zero premium, which is unacceptable. On the other hand, the expected value of this $\overline{P}$ is P, desirable if the system is to be 'fair'.

An alternative is sought, and that is to find an estimator of P that is never zero and whose expected value is equal to P. An obvious possibility worth trying is to form $\overline{P}$ by replacing τ in (1) by its maximum likelihood estimator m, the mean, given by

$$m=\frac{1}{n}\sum_{i=1}^{n}x_i, \qquad (2.2.4)$$

which leads to

$$\overline{P}=me^{-\frac{c}{m}}. \qquad (2.2.5)$$

Unless $m=0$, implying that each $x_i=0$ (and therefore hardly a case for reinsurance), $\overline{P}$ would be non-zero. Its expected value, since

$$d\ Prob(m<t) = \frac{n}{\tau} e^{-\frac{nt}{\tau}} \frac{(\frac{nt}{\tau})^{n-1}}{(n-1)!} dt, \tag{2.2.6}$$

is

$$E(\bar{P}) = \frac{1}{(n-1)!} \int_0^\infty (\frac{nt}{\tau})^n e^{-\frac{nt}{\tau} - \frac{C}{t}} dt. \tag{2.2.7}$$

This can be expressed in terms of modified Bessel functions [2] and does not look much like P given by (2.2.2). Numerical calculation suggests that $E(\overline{P})/P>1$, so that the choice of (2.2.5) to estimate the risk overstates the case.

J. Bather of the University of Sussex has suggested using for $\overline{P}$ the *minimum-variance unbiased estimator*

$$\bar{P} = m(1 - \frac{C}{nm})^n, \quad if\ C<nm,$$

$$=0, \quad if\ C \geq nm. \tag{2.2.8}$$

Using (2.2.6) we can check that $E(\overline{P})=P$ and the only disadvantage is that it could be zero. However, enjoyment of the minimum variance property means in practice that sample values lie closer to the mean than would be the case were any other unbiased estimator chosen, and this in turn implies that this estimator, among all other unbiased estimators, has the smallest chance of giving a zero $\overline{P}$.
The theory on which (2.2.8) depends is due to Rao and Blackwell. See [3] and [4] for the original work and, for example, [5] for a convenient and more accessible summary.

The form of $\overline{P}$ given by (2.2.8) shows why (2.2.5) overestimates P, for

$$(1 - \frac{C}{nm})^n < e^{-\frac{C}{m}}, \quad (n=1,2,3,...).$$

Thus the integrand in (2.2.7) exceeds

$$(\frac{nt}{\tau})^n e^{-\frac{nt}{\tau}} (1 - \frac{C}{nt})^n$$

for finite positive n and $t>C/n$, and it follows that the estimator (2.2.5) is biased, overstating the true value.

We look finally at another method of assessing the risk to the cedent. Let us suppose that he decides to use the mean m taken over the previous n years as priority. That is to say he decides to insure against a claim exceeding the average taken over the previous n years. In this case

$$P=\tau e^{-\frac{m}{\tau}} \tag{2.2.9}$$

is the risk in year $n+1$ against which he wishes to cover himself. The long-run value of P is its expectation taken with respect to the distribution of m, namely

$$E(P)=\frac{\tau}{(n-1)!}\left(\frac{n}{\tau}\right)^n\int_0^\infty e^{-\frac{m}{\tau}}m^{n-1}e^{-\frac{nm}{\tau}}\,dm$$

$$=\tau\left(\frac{n}{1+n}\right)^n. \tag{2.2.10}$$

The obvious analogue to (2.2.9) is to replace τ by m and to use the estimator

$$\overline{P} = me^{-1}, \tag{2.2.11}$$

but this has expected value

$$E(\overline{P})=\tau e^{-1}, \tag{2.2.12}$$

and since

$$lnE(P)=\ln\tau-nln\left(1+\frac{1}{n}\right)=\ln\tau-1+\frac{1}{2n}+...,$$

we have

$$E(P)>\tau e^{-1}=E(\overline{P}).$$

Thus (2.2.11) **underestimates** the expected risk. However, we can easily propose the alternative estimator

$$\bar{P}=m\left(\frac{n}{1+n}\right)^n, \tag{2.2.13}$$

which is unbiased, has the minimum variance property, and is not zero unless m is.

References

1. Vajda, S. (1951) Analytical studies in Stop-Loss reinsurance. Skandinavisk Aktuarietidskrift, 159-175.

2. Conolly, B.W. (1956) Unbiased premiums for Stop-Loss reinsurance. ibid., 127-134.

3. Rao, C.R. (1945) Information and accuracy attainable in estimation of statistical parameters. Bull. Calcutta Math. Soc.,37, 81-91.

4. Blackwell, D. (1957) Conditional expectation and unbiased sequential estimation. Ann. Math. Stat., 18, 105-110.

5. Mood, Alexander M. and Graybill, Franklin A.(1963) *Introduction to the theory of statistics*. 2nd edition. McGraw-Hill.

3

Games

The first of these two essays examines, using statistical and computational methods, certain features of the universally popular game of football: a rational method for the ranking of teams on the basis of their performance (not always unambiguous); testing a hypothesis that teams modulate performance according to performance so far; forecasting results, with particular reference to football pools. This study includes properties of Poisson processes and a less-known generalization, as well as other probability concepts. The second essay discusses, using graph theory, a special kind of game; a lively example for an algebra class.

3.1 FOOTBALL RESULTS

3.1.1 Rankings

In tournaments played according to the League System a player, or team, is given a certain number of points for winning, and another, smaller, number if the game ends in a draw. In the English Football League a win counts 3 points and a draw 1 point, but it is more usual for a win to be worth 2 points.

As examples we quote the semi-finals in Group A of the European Cup for the season 1991/92. Each team played every other twice, once at home, and once away. The results were:

Home Team	**Away Team**			
	S	**R**	**A**	**P**
S(ampdoria)	-	2:0	2:0	1:1
R(ed Star Belgrade)	1:3	-	3:2	1:0
A(nderlecht)	3:2	3:2	-	0:0
P(anathenaikos)	0:0	0:2	0:0	-

In terms of points for home and away games combined we obtain the following matrix (n_{ij}):

	S	**R**	**A**	**P**	Total
S	-	4	2	2	8
R	0	-	2	4	6
A	2	2	-	2	6
P	2	0	2	-	4

Two teams gained the same number of points. In such a case the ranking may be decided by the goals scored in the games between the relevant teams. In the present case, the games between **R** and **A** ended 3:2 and 2:3, and this is of no help. It has been suggested by Wei [1] that account should be taken of the strength of those teams against which success was achieved in the following way: multiply the n_{ij} of team $\mathbf{P}_i$ against team $\mathbf{P}_j$ $(i,j=1,\ldots,k)$ by the total

$$T_j=\sum_{i=1}^{k} n_{ji},$$

and rank according to

$$\sum_j n_{ij}T_j.$$

In this way a success against an opponent who proved to be weak during the tournament would have less weight than that achieved against a strong one.
In our example the ranking would become

S	4x6 + 2x6 + 2x4 = 44
R	0x6 + 2x6 + 4x4 = 28
A	2x8 + 2x6 + 2x4 = 36
P	2x8 + 0x6 + 2x6 = 28

The tie between **R** and **A** has been resolved, but one between **R** and **P** has been created. The games **R-P** and **P-R** ended 1:0 and 0:2, respectively, so that **R** must now be given preference. The resulting ranking **SARP** would not contradict the previous ranking. If we repeated the Wei procedure we should have

S-240, **A**-200, **R**-184, **P**-160,

again **SARP!**

What we have done in effect was to multiply the matrix (n_{ij}) by the transposed vector $(T_1,\ldots,T_k)^T$.

It can be proved that, by applying the Wei method repeatedly, the resulting ranking vectors (normalized, since they increase in magnitude from step to step) converge to an eigenvector of the largest eigenvalue of the matrix (n_{ij}). In the present case these quantities are:

largest eigenvalue: 5.76576369
eigenvector: $(1.58858790x,\ 1.14270862x, 1.29429395x, x)^T$.

Here x is a scale factor which can be set to unity. However, to see how closely the second-stage Wei iteration agrees with the limit we can put $x=160$. The resulting numbers, taken in ascending order of magnitude, are: 160x1.1427=182, 160x1.2943=207, 160x1.5886=254. Continuation of Wei iteration eight times gives the following approximation to the eigenvector up to the scale factor x:

$$(1.5845, 1.1384, 1.2922, 1).$$

3.1.2 Contagious distributions

Mr. Lancaster and Mr. York are the best of friends. In minor matters, such as the state of the economy, or a royal marriage, they invariably agree. But when it comes to football the red and the white rose can not call a truce. Luckily, in the 1991/2 season, both matches between Manchester United and Leeds United ended in 1:1 draws.

Before attending a match together, each of them has some idea regarding the probability of their favourite team scoring. However, during the game, their estimates tend to change, depending on the scores already obtained. At the end of the season, which saw Leeds first with 82 points and Manchester United runner-up with 78 points, they told me what their guesses had been and asked me what I, as a statistician, thought of it. I was rather dubious about the possibility of modelling their prejudices, but I did try one admittedly simplistic assumption.

Let the probability of a team scoring in the small time interval between t and $t+h$, *when the team has already scored n times*, be denoted by $\mu_n(t)h$. This means that the tendency to score depends on the length of time elapsed so far as well as on the number of goals. The number of goals scored in the time interval $(0,t)$ is known as a *non-negative, integer-valued random variable, and we shall denote it by N(t).* We shall be interested mainly in the probability $P_n(t)$ that $N(t)=n$ $(n \geq 0)$.

It satisfies the difference equations

$$P_0(t+h) = P_0(t)\{1-\mu_0(t)h\}$$

$$P_n(t+h) = P_{n-1}(t)\mu_{n-1}(t)h + P_n(t)\{1-\mu_n(t)h\}.$$

These can be rearranged as

$$\frac{P_0(t+h)-P_0(t)}{h}=-\mu_0(t)P_0(t)$$

$$\frac{P_n(t+h)-P_n(t)}{h}=P_{n-1}(t)\mu_{n-1}(t)-P_n(t)\mu_n(t). \qquad (3.1.1)$$

The P_0-equation is derived from the argument that no goals by time $t+h$ requires no goals in $(0,t)$ and no goal in $(t,t+h)$, which, under the supposed mechanism, are independent. The P_n-equation expresses the fact that n goals scored by time $t+h$ is the result either of n-1 by time t and one in $(t,t+h)$, or of n by time t and none in $(t,t+h)$. Our μ-mechanism, let it be noted, implies that the probability of more than one goal in the *small* interval $(t,t+h)$ is vanishingly small compared with the terms retained.

By letting $h->0$ we convert the difference equations into differential-difference equations as follows:

$$\frac{dP_0(t)}{dt}=-\mu_0(t)P_0(t)$$

$$\frac{dP_n}{dt}=\mu_{n-1}(t)P_{n-1}(t)-\mu_n(t)P_n(t),\ (n\geq 1). \qquad (3.1.2)$$

We must now make some assumption about the manner in which $\mu_n(t)$ depends on n and t, and we shall use the 'contagious' Eggenberger-Pólya form [2]

$$\mu_n(t)=\frac{\mu+an}{1+at},\quad (\mu,a>0).$$

The solution of the resulting differential-difference equation (see(3.1.3) is given by (3.1.4) subject to the obvious initial condition $P_0(0)=1$ (certainty that no goals are scored in a game of length zero),

$$\frac{dP_n(t)}{dt} = \frac{\mu+(n-1)a}{1+at}P_{n-1}(t) - \frac{\mu+na}{1+at}P_n(t) \qquad (3.1.3)$$

which holds for $n \geq 0$ by interpreting $P_{-1}=0$.

$$P_0(t) = (1+at)^{-\mu/a},$$

$$P_n(t) = \frac{\mu(\mu+a)...(\mu+(n-1)a)}{n!}\frac{t^n}{(1+at)^{n+\mu/a}}. \qquad (3.1.4)$$

The solution can be checked by substitution. The distribution has mean μt and variance $\mu t(1+at)$. With $a=0$ it is the Poisson distribution with mean rate μ.

We might mention as historical curiosities that the Poisson distribution was thought by L. von Bortkiewicz to describe the frequency of recruits to the Prussian army killed by horse kicks. The result (3.1.4) was published in [2] and applied to deaths by boiler explosions (also in Prussia), and to cases of scarlet fever and smallpox in Switzerland. It is understandable that this type of distribution should be called *contagious*. It is, of course, relevant to actuarial studies and in this context we refer the reader to the encyclopaedic reference [3]. Here we apply (3.1.4) to football results. The distribution of goals scored in League matches in the First Division during the 1991/92 season by Leeds United and Manchester United in their matches was as follows:

Leeds (Home)

0	1	2	3	4	5
2	9	4	4	1	1

Leeds (Away)

0	1	2	3	4	5	6
7	5	2	3	3	0	1

Manchester (Home)

0	1	2	3	4	5
2	11	4	2	1	1

Manchester (Away)

0	1	2	3	4	5	6
7	6	4	3	0	0	1

The means and variances of numbers of goals scored are:

Leeds(Home)	1.810	1.583	Manchester(Home)	1.619	1.474
Leeds(Away)	1.714	2.967	Manchester(Away)	1.381	2.140

In home games the variance was smaller than the mean. This negative contagion seems to give credibility to the widely held belief that, after having scored, a team concentrates on defence in preference to increasing their goal difference at the end of the season. Taking the length of a match as the unit of time we find from the results above the following estimates:

Leeds(H): $\mu=1.810$, $a=-0.125$
Manchester(A): $\mu=1.619$, $a=0.550$.

With these values for μ and a we give below a joint table of P_n and the corresponding theoretical goal frequencies in parentheses, which can be compared with the observations tabulated above.

	0	1	2	3	4	5	6	≥7
Leeds(H)	0.145	0.299	0.288	0.171	0.070	0.021	0.005	0
	(3.0)	(6.3)	(6.0)	(3.6)	(1.5)	(0.4)	(0.1)	0
Manch.(H)	0.183	0.326	0.274	0.144	0.054	0.015	0.003	0.001
	(3.8)	(6.8)	(5.8)	(3.0)	(1.1)	(0.3)	(0.1)	(0)
Leeds(A)	0.276	0.274	0.193	0.118	0.067	0.036	0.018	0.018
	(5.8)	(5.8)	(4.1)	(2.5)	(1.4)	(0.8)	(0.4)	(0.4)
Manch.(A)	0.333	0.296	0.185	0.099	0.048	0.022	0.010	0.007
	(7.0)	(6.2)	(3.9)	(2.1)	(1.0)	(0.5)	(0.2)	(0.1)

The best detailed agreement with observation is Manchester (A). However, it seems unwise to draw conclusions about contagion from such a small, 'one-off' sample, especially as the validity of the Pólya model for negative a, which occurs in both sets of home game results, is dubious. (Note that $\mu_n(t)$ is defined here only for $a>0$.) It has also been proposed that there is a possible tendency in home games for teams to concentrate on defence after having scored. This would suggest a higher frequency of lower scores than in away games, but this is not very evident in these results. So, regretfully, our attempt to import a contagious distribution as a model for football results can not be considered to have been entirely successful. But it was worth a try and we hope that the reader agrees. Further attention is given to the problem in the next part.

3.1.3 Winning the Pools?

"You statisticians are lucky", ventured my neighbour at a dinner party one evening. "How so?", I asked. "Well, to begin, you must be terribly rich."
"Wait a minute", I said. "I shall have to ask you to change places with me. I am a little deaf in one ear today and I absolutely can not believe what I think I heard you say." After a small commotion, which temporarily disturbed the urbanity of the proceedings, my neighbour and I managed to exchange seats. "Now", I said, "would you mind repeating what you said about statisticians being rich?" "Yes, that is correct" she replied. "My husband is also rich. He sells scrap metal, but he has to be at it from morning to night, almost seven days a week. I hardly ever see him. You statisticians can get rich just by sitting at your desks." "I am not really a statistician", I protested:"I am also rather poor. So maybe, for my benefit, you could let me have some idea how it is that statisticians can enrich themselves with so little trouble."

"It's the gambling", explained my neighbour. "You know all the tricks they get up to at Monte Carlo. You sit at the tables with your little books, observe what is going on, do a few calculations - and Hey!Presto!, you've made a fortune." "I think you are a romantic", I said. "Every student of probability knows that it is the casino owners who are the winners at that game. If they happen to be statisticians themselves as well...." I left the sentence unfinished, a little worried about trapping myself in my own postprandial assertiveness.

"You don't have to go as far as Monte Carlo", continued my companion, ignoring my remark and sipping delicately at her wineglass which I noticed was empty. "There are rich pickings nearer home for those versed in the probability craft." "And what are they, pray?", I queried. "Come off it", she replied, a trifle vulgarly I thought, for an elegant lady: but then she had married a scrap merchant. "There's the racing." "You must read", I said, "a little book my friend and I have put together. That shows that horses are for pleasure rather than profit." "Well", she replied, undaunted, "there's the pools." "What", I exclaimed, "You mean the football pools?" "Exactly" she replied. "And I think from the cunning expression I notice spreading across your face that I have hit the nail on the head. Come on now: confess!"

Well, it was true that I had given football pools a little thought, but rather for the alleviation of boredom than in pursuit of profit, which I soon realised to be vain. And this, in brief, is what I told my neighbour.

It is common knowledge that to win a football pool it is necessary to forecast correctly the outcome of a set of games between teams, scheduled to take place in the near future. There are variations on this theme but it is enough to stay with the simplest. So suppose the objective is to forecast the results of 10 matches between 20 teams. A match between teams i and j may result in a win for either i or j, depending on which scores the greater number of goals, or it may be a draw if each team scores the same number of goals. One of i and j is the 'home' team, and the other is the 'away' team, meaning that one plays on its home ground, thought by many experts to enhance the chances of victory. The opponent is then, of course, not at home, and although many fervent supporters may accompany the team it is generally felt that they are, in this position, at a moral disadvantage, at the very least. Typically, one point is awarded for a correct forecast of a 'home' win, two points for an 'away' win, and three points for a draw. The aspirant who chooses this road to fortune has effectively to make a correct, or almost correct, forecast of the outcome of all ten matches, and since there are three possibilities for each, the winning entry is one of $3^{10}=59049$ possible sets of ten. Each forecast costs the forecaster (or "punter") an entrance fee, so if he is to be sure of winning, his outlay will be considerable, and the larger the number of other punters who have adopted a similar strategy, the broader will be the smiles wreathing the faces of the pools promoters as they scoop up the entrance fees, especially when the prize is a single one that has to be divided among multiple winners. Guess who is going to be the net winner week after week, and why didn't you think of it first?

Unless - and it is a very big unless - some guidance can be found concerning the outcome of a match. Based on 'form", that is, a study of teams' records during the season, it seems plausible to hope that an assessment of probability of winning might be made, and that is what I am about to discuss. The simplest model one might think of is that the goals scored by team i in *any* match, whether at home or away, and irrespective of opponent, is a simple Poisson process with mean λ_i. Then a match between teams i and j will result in a win for team i if team i scores more goals than team j. The probability that a team with mean λ scores r goals, on the assumption of a Poisson process, is $\exp(-\lambda)\lambda^r/r!$. Thus, the probability that team i, with mean λ_i, scores $n+r$ goals and, **independently,** that team j scores just r is the product of the individual probabilities, namely,

$$e^{-\lambda_i}\frac{(\lambda_i)^{n+r}}{(n+r)!}e^{-\lambda_j}\frac{(\lambda_j)^r}{r!}.$$

Then, the probability that,in a game between i and j, team i scores n more goals than team j, is this expression summed over all possible goal scores of team j, that is r=0,1,2..., giving the probability

$$e^{-\lambda_i-\lambda_j}\sum_{r\geq 0}\frac{(\lambda_i)^{n+r}(\lambda_j)^r}{(n+r)!r!}.$$

Finally, the probability W_{ij} that team i **wins** is obtained by summing this expression over all n compatible with a win, giving

$$W_{ij}=e^{-(\lambda_i+\lambda_j)}\sum_{n\geq 1}\sum_{r\geq 0}\frac{(\lambda_i)^{(n+r)}}{(n+r)!}\frac{(\lambda_j)^r}{r!}. \qquad (3.1.5)$$

If n=0 the result is a draw, and so the corresponding probability D_{ij} is

$$D_{ij}=e^{-(\lambda_i+\lambda_j)}\sum_{r\geq 0}\frac{(\lambda_i\lambda_j)^r}{r!r!}, \qquad (3.1.6)$$

and the probability L_{ij} that team i loses is

$$L_{ij}=1-W_{ij}-D_{ij}. \qquad (3.1.7)$$

by total probability.

And I might as well mention that the probability of a draw is usually a good deal less than that of a win or a loss.

For the mental improvement of my dinner companion I mentioned that these probabilities can all be expressed in terms of modified Bessel functions of the first kind.

See the next three expressions.

$$W_{ij}=e^{-(\lambda_i+\lambda_j)}\sum_{n\geq 1}\left(\frac{\lambda_i}{\lambda_j}\right)^{n/2}I_n[2\sqrt{(\lambda_i\lambda_j)}], \tag{3.1.8}$$

$$D_{ij}=e^{-(\lambda_i+\lambda_j)}I_0[2\sqrt{(\lambda_i\lambda_j)}], \tag{3.1.9}$$

$$L_{ij}=e^{-(\lambda_i+\lambda_j)}\sum_{n\geq 1}\left(\frac{\lambda_j}{\lambda_i}\right)^{n/2}I_n[2\sqrt{(\lambda_i\lambda_j)}]. \tag{3.1.10}$$

These probabilities, which *may* provide useful guidance to a would-be forecaster, can be evaluated only if λ_i and λ_j are known. So that is the next question.

To find these, fortune-hunters must study form: that is, they must collect data and process it appropriately.

"Like a statistician", my companion interrupted triumphantly. "It's beginning to sound as though I was right about statisticians and wealth." "Nothing of the kind", I snorted. "Just pay attention a little longer and you will see how wrong you are."

The publication of the first games of the season will be a starting point. In this discussion, where 20 teams are in question, they take the form of a set $(n_i{:}i=1..20)$ of goals scored in the first matches by each of the 20 teams, and, of course, n_i is a non-negative integer. It is known that λ_i is estimated by the average number of goals scored by team i taken over a sequence of, say, m matches. Thus λ_i is estimated by $(n_1+n_2+\ldots+n_m)/m$. Using this estimate the probable outcome of the matches in week $m+1$ can be assessed, the coupon filled-in and a pleasant time passed awaiting the arrival of the winnings cheque! At this point my companion announced that she had lost me. "Could you please give me an example?" "There's no time now" I replied. "You will have to meet me in my office tomorrow, when I shall explain the *modus operandi*.

The next day came, together with my previous evening's acquaintance. I showed her the following Table 3.1.1, which gives the goals scored by 20 teams on ten successive weeks. Before doing so I had to confess that these were not real results, but random Poisson variables generated on my computer using prescribed λ-values that I would reveal later. I intended to use these as the basis of a numerical experiment. The method and programs are described briefly in the Notes.

"The first task", I went on, "is to estimate λ_{mn} for team $m(=1..20)$ based on their goal scores for weeks 1..n. This produces Table 3.1.2 below.

Note: The true values of λ_m are given in the final column and it is clear that the estimates based on the goal scores week by week are, at the best, very sketchy: a gloomy prospect for the forecasting operation!

The next task is to 'arrange' 10 fictional sets of different weekly matches between the 20 teams, taken in pairs. There are 190 possibilities and the choice of pairings is quite arbitrary. It is given, together with the result in favour of the home team, in Table 3.1.3. In the fourth column of each of the mini-tableaux is given the prediction based on the probabilities calculated from (3.1.5), (3.1.6) and (3.1.7). [The Notes in 3.1.5 contain a Pascal program for the procedure used.]

Table 3.1.1 Goals scored by 20 teams on 10 successive weeks

Week/ Team	1	2	3	4	5	6	7	8	9	10
1	0	4	0	3	2	6	4	3	4	7
2	1	3	5	3	3	2	4	3	2	1
3	2	3	4	2	1	0	3	2	0	2
4	1	2	4	3	2	2	1	2	4	1
5	8	1	4	2	1	2	2	3	2	2
6	1	2	0	4	4	4	3	2	1	2
7	1	2	1	1	1	2	3	4	1	9
8	1	3	2	3	1	3	4	2	4	4
9	1	2	2	0	2	2	3	4	3	0
10	3	5	2	0	5	0	1	3	4	2
11	2	6	0	1	2	2	0	3	2	2
12	2	2	0	2	6	2	2	6	2	2
13	4	2	3	1	3	4	5	3	0	3
14	4	6	3	3	3	0	5	2	5	1
15	5	6	4	7	3	2	9	0	3	5
16	3	1	2	1	2	3	2	3	2	2
17	1	3	2	6	5	3	3	2	4	0
18	1	3	2	3	2	2	2	1	4	4
19	1	3	3	3	2	3	3	3	3	3
20	5	2	6	4	6	3	2	3	4	3

Table 3.1.2 λ-estimates

Week/ Team	1	2	3	4	5	6	7	8	9	10	true
1	0	2	1.33	1.75	1.8	2.5	2.71	2.75	2.89	3.3	2.5
2	1	2	3	3	3	2.83	3	3	2.89	2.7	2.6
3	2	2.5	3	2.75	2.4	2	2.14	2.12	1.89	1.9	2.7
4	1	1.5	2.33	2.5	2.4	2.33	2.14	2.12	1.89	1.9	2.7
5	8	4.5	4.33	3.75	3.2	3	2.86	2.88	2.78	2.7	2.9
6	1	1.5	1	1.75	2.2	2.5	2.57	2.5	2.33	2.3	3
7	1	1.5	1.33	1.25	1.2	1.33	1.57	1.88	1.78	2.5	3
8	1	2	2	2.25	2	2.17	2.43	2.38	2.56	2.7	3
9	1	1.5	1.67	1.25	1.4	1.5	1.71	2	2.11	1.9	3
10	3	4	3.33	2.5	3	2.5	2.29	2.38	2.56	2.1	3
11	2	4	2.67	2.25	2.2	2.17	1.86	2	2	2	3
12	2	2	1.33	1.5	2.4	2.33	2.25	2.75	2.67	2.6	3
13	4	3	3	2.5	2.6	2.83	3.14	3.12	2.78	2.8	3
14	4	5	4.33	4	3.8	3.17	3.43	3.25	3.44	3.2	3
15	5	5.5	5	5.5	5	4.5	5.14	4.5	4.33	4.4	3
16	3	2	2	1.75	1.8	2	2	2.12	2.11	2.1	3.1
17	1	2	2	3	3.4	3.33	3.29	3.12	3.22	2.9	3.2
18	1	2	2	2.25	2.2	2.17	2.14	2	2.22	2.4	3.3
19	1	2	2.33	2.5	2.4	2.5	2.57	2.62	2.67	2.7	3.4
20	5	3.5	4.33	4.25	4.6	4.33	4	3.88	3.89	3.8	3.5

Table 3.1.3 The matches, the results and the predictions

	Week	1			Week	2			Week	3	
i	j	Act	Pred	i	j	Act	Pred	i	j	Act	Pred
1	2	l	-	1	4	w	l	1	6	d	w
3	4	w	-	3	6	w	w	3	8	w	w
5	6	w	-	5	8	l	w	5	10	w	w
7	8	d	-	7	10	l	l	7	12	w	l
9	10	l	-	9	12	d	l	9	14	l	l
11	12	d	-	11	14	d	l	11	16	l	w
13	14	d	-	13	16	w	w	13	18	w	w
15	16	w	-	15	18	w	w	15	20	l	w
17	18	d	-	17	20	w	l	17	2	l	d
19	20	l	-	19	2	d	w/l	19	4	l	w

	Week	4			Week	5			Week	6	
i	j	Act	Pred	i	j	Act	Pred	i	j	Act	Pred
1	8	d	l	1	10	l	l	1	12	w	l
3	10	l	l	3	12	l	w	3	14	d	l
5	12	d	w	5	14	l	l	5	16	l	w
7	14	l	l	7	16	l	l	7	18	d	l
9	16	l	l	9	18	d	l	9	20	l	l
11	18	l	w	11	20	l	l	11	2	d	l
13	20	l	w	13	2	d	l	13	4	w	w
15	2	w	w	15	4	w	w	15	6	l	w
17	4	w	l	17	6	w	w	17	8	d	w
19	6	l	w	19	8	w	w	19	10	w	l

	Week	7			Week	8			Week	9	
i	j	Act	Pred	i	j	Act	Pred	i	j	Act	Pred
1	14	l	l	1	16	d	w	1	18	d	w
3	16	w	w/l	3	18	w	w/l	3	20	l	l
5	18	d	w	5	20	d	l	5	2	d	l
7	20	w	l	7	2	w	l	7	4	l	l
9	2	l	l	9	4	w	l	9	6	w	l
11	4	l	l	11	6	w	l	11	8	l	l
13	6	w	w	13	8	w	w	13	10	l	w
15	8	w	w	15	10	l	w	15	12	w	w
17	10	w	w	17	12	l	w	17	14	l	l
19	12	w	w	19	14	w	l	19	16	w	w

	Week	10	
i	j	Act	Pred
1	20	w	l
3	2	w	l
5	4	w	w
7	6	w	l
9	8	l	l
11	10	d	l
13	12	w	w
15	14	w	w
17	16	l	w
19	18	l	w

No prediction can be given for Week 1 since no prior value of the λ's is available. Now take Week 2. The matches are between team pairs i versus j as follows: 1 v. 4, 3 v.6,...,19 v.2, and the result in favour of team i, the 'home' team, is obtained from Table 3.1.1 and given in the third column. For example, in Week 2, Team 1 scores 4 goals and Team 2 scores 3. Thus, the result is w(in), a home win for Team 1. Next, in the match 3 v. 6, Team 3 scores 3 goals and Team 6 scores 2, giving a home win for Team 3; and so on. The fourth column of predicted results is obtained by using the appropriate λ's from the preceding week in Table 3.1.2. From Week 1, Table 3.1.2, we get $\lambda_1=0$, $\lambda_4=1, \lambda_3=2, \lambda_6=1$. For the first two matches of Week 2 we then calculate

$W_{14}=0.00$, $D_{14}=0.37$, $L_{14}=0.63$,
$W_{36}=0.61$, $D_{36}=0.21$, $L_{36}=0.18$,

and so, since L_{14} is the largest probability in the (1,4) game and W_{36} the largest in the (3,6) game, we predict l (lose) for the first match and w (win) for the second. Incidentally, when $\lambda=0$, the computer program requires that a very small non-zero value be used; otherwise the computer slips in an error message and sulks. The occasional entries w/l mean that $W_{ij}=L_{ij}$ and that both exceed D_{ij}, that is, win and lose have equal probabilities while the probability of a draw is the smallest. Almost any prediction seems feasible!

As both the reader has, and my client had, by now realised, the prediction performance based on scores obtained from the Poisson assumption is far from encouraging. In the 9 successive weeks from 2 to 10 we get at most 42 correct predictions - less than 50%.

"This is all very well", said my friend, "but I would be more convinced by actual game results. All this computer randomizing nonsense is highly doubtful. I might as well go to a fortune-teller, or a medium."

"As it happens", I replied, "I have just received five weeks' results of matches in the 1991 season betweeen pairs of teams drawn from the 22 clubs in the First Division of the League as it then was. I have also a record of the goals scored by each team in each of the five weeks. If you will be so kind as to return tomorrow I shall be able to tell you how similar exercises with real data turn out." She agreed, and when the next day came I explained that I had experimented both with the Poisson assumption **and** with the additional assumption that goals scored by a team in a match could be modelled by the Pólya process introduced in Part 3.1.2.

The goal data are given in Table 3.1.4. I decided to use all five weeks' results for each team to calculate an overall λ parameter for forecasting results on the Poisson model, and likewise to estimate the μ and a parameters for Pólya forecasting. This, in principle, puts the forecaster in a better position than if he uses just the accumulated results up to week n to forecast results for week $n+1$. The estimated values are given in the final columns of Table 3.1.4. λ and μ are estimated by the mean of five weeks. Thus, Team 1 scored 9 goals altogether, giving $\lambda=\mu=9/5=1.8$. a is estimated from the variance $v=\mu(1+a)$. For Team 1, $v=1.36$, so the a-estimate is -0.24.

Table 3.1.4 Goals scored by 22 teams in 5 weeks of games

Team	Week 1	2	3	4	5	$\lambda=\mu$	a
1	1	1	2	1	4	1.8	-0.24
2	0	3	1	2	1	1.4	-0.26
3	2	3	4	1	0	2.0	0
4	0	1	0	2	1	0.8	-0.3
5	1	2	2	2	1	1.6	-0.85
6	1	0	1	2	1	1	-0.6
7	0	1	1	3	1	1.2	-0.2
8	2	0	3	2	1	1.6	-0.35
9	0	0	1	0	2	0.6	0.07
10	1	3	1	0	0	1	0
11	2	0	1	3	1	1.4	-0.26
12	2	2	0	0	2	1.2	-0.2
13	2	4	3	1	4	2.8	-0.51
14	0	0	2	1	0	0.6	0.07
15	1	2	1	2	1	1.4	-0.83
16	1	1	1	2	0	1	-0.6
17	2	0	1	1	2	1.2	-0.67
18	2	1	1	2	1	1.4	-0.83
19	2	2	1	2	0	1.4	-0.54
20	3	1	1	5	2	2.4	-0.07
21	0	0	0	1	1	0.4	-0.4
22	2	2	1	3	2	2	-0.8

Table 3.1.5 below gives the actual results for Weeks 4 and 5. For each week are given the contending teams i and j, i being the home team; the actual result (A) as L(ose), D(raw), W(in); the predicted result given by the largest probability calculated on the Poisson model (Poi), and on the Pólya model(Pól). The probabilities are calculated as explained above in connection with the simulation exercise using the program fpool listed in the Notes. For the Pólya case the analogous formulae are used and calculated by the program fpol, also listed in the Notes. This latter program needs a little attention since it sometimes returns an error message, but the results are almost identical with the Poisson prediction where they can be compared. In Week 4 the Poisson model correctly predicts the result in 6 cases out of 11 and, where a result is obtained, the Pólya prediction is identical. In Week 5 the Poisson model again gets 6 right out of 11, with one ambiguous W/L. The Pólya model gets one prediction correct although Poisson gets it wrong (the 4-14 match). The conclusion ventured earlier, although based on simulation, and omitting the Pólya method, is difficult to escape. This way is not a direct path to riches for anyone, and certainly not for statisticians. Better to stick to scrap-metal dealing. In fact a mere look at team averages might be as reliable a guide to results as these more sophisticated models, and it seems on the evidence that there is not much to

choose between a Poisson and a Pólya model. But it is all good fun!

Table 3.1.5 Actual and predicted results for weeks 4 and 5

		Week	4				Week	5	
i	j	A	Poi	Pol	i	j	A	Poi	Pol
1	4	L	W	W	3	7	L	W	W
2	20	L	L	L	4	14	W	D	W
6	5	D	L	-	5	1	L	L	L
7	10	W	W	W	8	2	D	W	W
11	12	W	W	W	9	15	W	L	-
14	8	L	L	L	10	18	L	L	-
15	17	W	W	-	12	21	W	W	W
16	19	D	L	-	13	22	W	W	-
18	13	W	L	-	17	6	W	W	-
21	3	D	L	L	19	11	L	W/L	W
22	9	W	W	W	20	16	W	W	W

3.1.4 And a postscript on syndicates

The next day I received a telephone call. "Mrs.X told me", the voice said, "that you are pessimistic about football pools as a reliable get-rich-quick mechanism."
"True", I replied. "Then what about the story I have just heard", the voice continued, "about the syndicate which recently gained a £1,000,000 win, and, moreover, the forty-fifth since they began working together? How do you explain that?". I invited the caller round and gave him a short lecture as follows.

At a very early stage in probability classes one is taught about **Bernoulli Trials**. These, roughly speaking, are experiments with just two possible outcomes:success (S), and failure (F). When a sequence of such experiments, or trials, is carried out the rule is that the outcome of each trial is independent of what happened in the preceding trials. The probability of success is usually denoted by p, and of failure by q, and since there are only these two possibilities $p+q=1$. "I shall ask you", I declared, "to accept that the probability that n successes, without regard to order, are obtained in a sequence of N trials is given by

$$P_n = \binom{N}{n} p^n q^{N-n},$$

where

$$\binom{N}{n}$$

is shorthand for the binomial coefficient

$$\frac{N(N-1)...(N-n+1)}{1.2.3...n}.$$

When you fill in a pool coupon, among the many variations proposed by benevolent operators is the basic one of asking the punter to forecast correctly the outcome of a set of, say, ten matches. An outcome is either a win or loss for the home team, or a draw. In order to arrive finally at helpful orders of numerical magnitude I am going to ask you to accept, although we now know well that it is not true, that each of these three possibilities has the same probability, namely 1/3. Then the probability that you get a whole line right is

$$(1/3)^{10} = 0.00001694,$$

and that is pretty small.[1] No wonder that people tend to fill their coupons with multiple entries! So let us suppose that I decide to submit ten different lines on my coupon each week. The probability that at least one line is correct is the probability that not all are wrong, and that is $1\text{-}q^{10}$. This is 0.00016939 - just about ten times the probability of getting a single line right, and if you are mathematically inquisitive you will easily see why this is the case when success probability is extremely small. "Yes,I see", said my interlocutor, "but what would be the chance of at least one correct line if I submit ten lines a week throughout the whole season?" "They can be the same ten lines each week", I began, "provided that they are all different in one week. But let us suppose that a season consists of 26 weeks. Then the probability that none of the 260 lines submitted (and at what cost?) is correct is q^{260}, and the probability that at least one line is correct is

$$1\text{-}q^{260} = 0.00439475.$$

Still not very encouraging!" "But suppose", interrupted my interlocutor, his excitement plainly rising, "I get up a syndicate of 50 friends to join me, making sure that all 500 lines submitted each week are different." "The probability that at least one line is

[1]In fact, the probability of a draw is smaller than that of a win. We take the probabilities equal to simpli[fy] the example.

correct in any given week", I said, "is $1\text{-}q^{500}=0.00843430$: some improvement over when you were doing it alone, but if your friends continue for a whole season, the probability rises to

$$1\text{-}q^{26*500} = 0.19765923$$

that, between you, you will get at least one correct line. That is now virtually 20%, or 4:1 against. If each syndicate member could pay an entry fee of 10p per line, then his outlay for the season would be £26, and if, as in the case you started with the prize were £1,000,000, that would be a return of £20,000 to each member, and that's not such a bad prospect at these odds, is it?". I wonder!

3.1.5 Notes

Listings follow of the three Pascal programs used in this study:

1. poisson; 2.fpool; 3. fpol.

1.**Poisson** This generates pseudorandom Poisson variates N with point probability $p_n=\exp(-\lambda)\lambda^n/n!$. The method is to make a table of p_n and to store it. Then a pseudo-uniform on (0,1) is generated (here using a handy method due to Wichmann and Hill[4]). $N=0$ is returned if $u<p_0$, and $N=n$ if $p_{n-1}\leq u<p_n$, $(n\geq 1)$.

2.**fpool** This generates the probability of win, draw or lose in a game between two teams, assuming independent goal-scoring Poisson mechanisms for each with parameters λ and μ. The formulae are the summations in (3.1.5), (3.1.6) and (3.1.7). The summation with respect to n is carried out first.

3. **fpol** This is like **fpool**, but the basis is two independent Pólya processes for the goal scoring mechanisms. The analogues to (1), (2) and (3) of 3.1.3 are used for the calculations and the method of organization is similar. The basic probabilities are given at (3.1.4). The program as shown needs a little work since it occasionally gives rise to overflow. Two parameters have to be entered for each of the opposing teams.

```
program poisson(input, output); uses printer;
label 1;
const mx=171;my=172;mz=170;m1=177;m2=176;m3=178;c1=2;c2=35;
c3=63;dx=30269;dy=30307;dz=30323;one=1;
var la,u,mn,sd:real;x,y,z,m,k,s:integer;
p:array[0..20] of real;q:array[0..20] of integer;
lst:text;
procedure wichill;
var x1,y1,z1,x2,y2,z2,n:integer;
procedure modd(var a:integer;b,c:integer);
begin a:=b-(b div c)*c end;
begin modd(x1,x,m1);x2:=x div m1; x:=mx*x1-c1*x2;
modd(y1,y,m2);y2:=y div m2; y:=my*y1-c2*y2;
```

```
modd(z1,z,m3);z2:=z div m3;z:=mz*z1-c3*z2;

if x<0 then x:=x+dx;
if y<0 then y:=y+dy;
if z<0 then z:=z+dz;
u:=x/dx+y/dy+z/dz;
while u>one do  u:=u-one;
end;{wichill}

procedure table;
var x:real;n:integer;
begin x:=exp(-la);
p[0]:=x;
for n:=1 to 20 do
begin x:=x*la/n;
p[n]:=p[n-1]+x;
end;
end;{table}

begin {program begins}
writeln('input x,y,z integer input seeds');
writeln('typically, small positive  integers');
read(x,y,z);writeln(x,'  ',y ,'  ',z);
write('input lambda');read(la);writeln(lst,'lambda= ',          la:1: 4);
writeln(lst);
table;writeln(lst,'table finished');writeln(lst);
for m:=1 to 100 do
begin wichill;
k:=0;
1: if u<p[k] then q[m]:=k else begin k:=k+1;goto 1; end;
end ;
for m:=1 to 10 do begin writeln(lst,m:4,'  ',q[m]:2) end;
s:=0;sd:=0;
for m:=1 to 100 do begin s:=s+q[m];sd:=sd+q[m]*q[m] end;
mn:=s/100;sd:=sd/100-mn*mn;
writeln('mn,sd= ',mn:1:4,'  ',sd:1:4);
writeln(lst, x,'  ',y,'  ',z);writeln(lst);writeln(lst);
end.{poisson}

program fpol(input, output);uses printer;
const eps=0.000001;
var mu1,a1,mu2,a2,l,m,u,v,w,x,t,xn,WIN,DRAW,LOSE:real;
r:integer;
```

```
        begin writeln('input mu1,a1,mu2,a2');
        readln(mu1,a1,mu2,a2);
        writeln(lst,'mu1= ',mu1:2:4,'  ','a1= ',a1:2:4);
        writeln(lst,'mu2= ',mu2:2:4,'  ','a2= ',a2:2:4);
        l:=mu1*(ln(1+a1))/a1;l:=exp(-l);
        m:=mu2*(ln(1+a2))/a2;m:=exp(-m);
        u:=1-l;v:=m;r:=1;x:=0;
        w:=u*v;xn:=w;t:=l;
        while abs(1-x/xn)>eps do
  begin x:=xn;v:=v*(mu2+(r-1)*a2)/(r*(1+a2));
  t:=t*(mu1+(r-1)*a1)/(r*(1+a1));
  u:=u-t;w:=v*u;xn:=x+w;
  r:=r+1
  end;
        writeln(lst,'Probwin=      ',xn:1:5);WIN:=xn;
        u:=l;v:=m;r:=1;w:=u*v;x:=0;xn:=w;
        writeln(lst);writeln(lst);
        writeln(lst)
        end.


        program fpool(input, output);uses printer;
        const eps=0.000001;
        var mu1,a1,mu2,a2,l,m,u,v,w,x,t,xn,WIN,DRAW,LOSE:real;
  r:integer;
        begin writeln('input mu1,a1,mu2,a2');
        readln(mu1,a1,mu2,a2);
        writeln(lst,'mu1= ',mu1:2:4,'  ','a1= ',a1:2:4);
        writeln(lst,'mu2= ',mu2:2:4,'  ','a2= ',a2:2:4);
        l:=mu1*(ln(1+a1))/a1;l:=exp(-l);
        m:=mu2*(ln(1+a2))/a2;m:=exp(-m);
        u:=1-l;v:=m;r:=1;x:=0;
        w:=u*v;xn:=w;t:=l;
        while abs(1-x/xn)>eps do
        begin x:=xn;v:=v*(mu2+(r-1)*a2)/(r*(1+a2));
        t:=t*(mu1+(r-1)*a1)/(r*(1+a1));
        u:=u-t;w:=v*u;xn:=x+w;
        r:=r+1
        end;
        writeln(lst,'Probwin=      ',xn:1:5);WIN:=xn;
        u:=l;v:=m;r:=1;w:=u*v;x:=0;xn:=w;
        while abs(1-x/xn)>eps do
```

```
begin x:=xn;u:=u*(mu1+(r-1)*a1)/(r*(1+a1));
     v:=v*(mu2+(r-1)*a2)/(r*(1+a2));
     w:=u*v;xn:=x+w;r:=r+1
end;
   writeln(lst,'Probdraw= ',xn:1:5);DRAW:=xn;
   LOSE:=1-WIN-DRAW;
   writeln(lst,'Problose= ',LOSE:1:5);
   writeln(lst);writeln(lst);
   writeln(lst)
   end.
```

References

1. Wei,T.H. (1952) The algebraic foundation of ranking theory. Thesis, Cambridge University.
2. Eggenberger, F. and Pólya, G. (1923). Über die Statistik verketteter Vorgänge. Zeitschrift für angewandte Mathematik und Mechanik. 3, 279-289.
3. Seal, H.L. (1969). *Stochastic theory of a risk business*, Wiley. See, in particular, Chapter 2.
4. Wichmann, B.A. and Hill, I.D. (1985) An efficient and portable pseudorandom number generator. pp. 238-242 in Applied Statistics Algorithms. Eds. P. Griffiths and I. D. Hill. Ellis Horwood, Chichester. (This algorithm was originally published in Applied Statistics.)

3.2 A CLASS OF GAME

Jack and Jill play the following game: from two piles of counters, with sizes m and n, respectively, and $m>n$, the players remove, alternately, any number of counters, but all from the same pile. The player who takes the last counter **is the winner**. Jill makes the first move. She is clever, and sees that she can win this game by using a **winning strategy**. Whenever the two piles are of unequal sizes, as they were to begin with, she makes them equal. Finally she will make both sides zero, thereby winning. To be certain that this strategy is winning it must possess the following properties:

(i) From a position with equal sizes the opponent can only produce one with unequal sizes;he can not make the sizes equal.

(ii) From a position with unequal sizes it is always possible to make them equal.

To obtain more insight we describe the possible moves by a graph whose vertices correspond to the possible positions during play, while the directed arcs describe moves which change a position into another. In Figure 3.2.1 we have chosen the initial sizes to be 2 and 3 and it will be clear how to construct a graph for larger sizes.

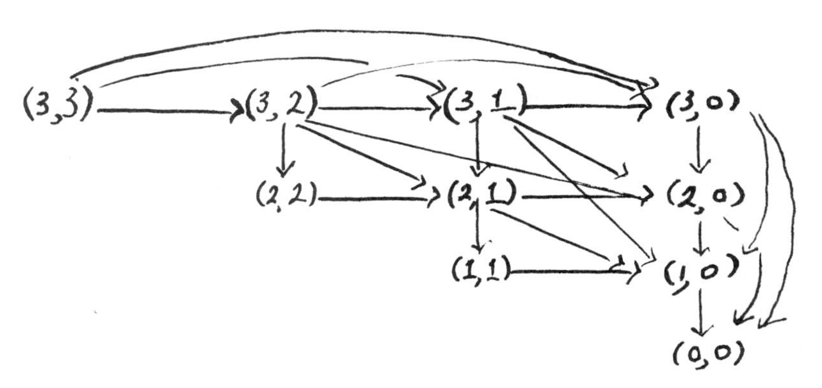

From any one vertex (0,0), (1,1) or (2,2) it is not possible to move directly to any other of these vertices, but from any other position one of these vertices can be reached. Such a set of vertices with the property that there is no possible transition from any one of them to another, but also such that at least one of them can be reached from any vertex not in the set, is called a **kernel**.

Imagine now that we have a directed graph with a kernel, with a single starting point and a single terminal point. Then we can play a game with the following rules: two players move alternately on successive arcs, and the player who reaches the end point wins. Clearly, the winning strategy consists in always moving to a vertex in the kernel. The player who moves first can use this strategy provided that the starting point is not in the kernel; if it is, the second player can use the winning strategy, while the first player can not. Thus, to play the game it is useful to find the kernel of a graph, if it exists. Not every graph has a kernel; if the graph contains a loop, or a circular path, then no kernel can exist.

The game with which we started is a special case of the game called NIM [1]. In general, Nim can be played with any number of piles. As a matter of fact, it is usually played with the rule that the player who takes the last counter loses; but the analysis of this version is more complex and we chose our version for simplicity, yet of mathematical interest. It would be hopeless to try to draw a graph even for a three-pile game with the object of finding the winning strategy by discovering the kernel. But this is not necessary because the label of any position can be ascertained directly in a rather elegant way.

Write the pile sizes in the binary scale, one below the other. For instance, if the pile sizes are 1,2,5,8, write

$$\begin{array}{r} 1 \\ 10 \\ 101 \\ \underline{1000} \\ \underline{1110} \end{array}$$

Compute the **Nim-sum** of these numbers by writing into each column of the sum, 0 if 1 appears an even number of times in that column, and 1 otherwise. In this example we obtain 1110. This is the binary equivalent of 14, but this interpretation is irrelevant. What matters is that if the Nim-sum is 00...0 (i.e. zero) then we have a position corresponding to a vertex in the kernel of the graph.

To see this, observe that if we change any one of the sizes, the Nim-sum 0 will change, while, on the other hand, if the Nim-sum is not 0, and we Nim-add it to (one of) the largest pile size(s), in our example

$$\begin{array}{r} 1000 \\ \underline{1110} \\ 110 \end{array}$$

we obtain the size to which that pile should be reduced. We obtain

$$\begin{array}{r} 1 \\ 10 \\ 101 \\ \underline{110} \\ 000 \end{array}$$

Therefore, the winning strategy consists of producing at each step the Nim-sum 0.

The reader will be easily convinced that this works in Figure 3.2.1 if labels are attached to the vertices as described below. Of course, for two sizes, a Nim-sum 0 is obtained only when the two sizes are equal.

We now exhibit a method for finding the kernel in the graph of Figure 3.2.2.

Figure 3.2.2

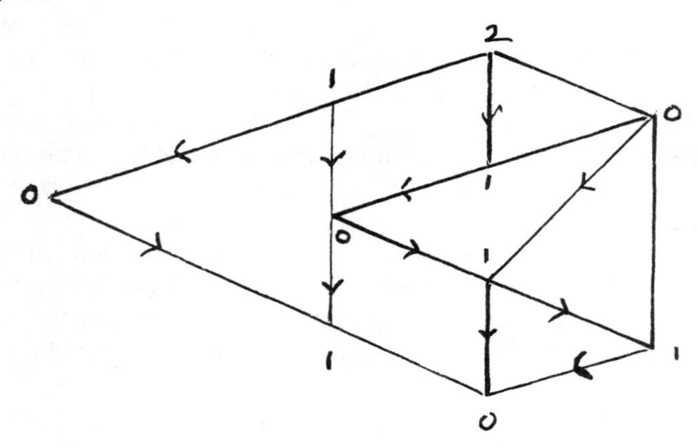

We label the vertices in the following manner. The label of the end-point is 0. Then we label at each step a vertex from which arcs lead only to vertices already labelled, and attach to this vertex, as its label, the smallest non-negative integer which is not in the set of those labels whose vertices can be reached from the point to be labelled. It is easily seen that the vertices labelled 0 form the kernel of the graph: a vertex labelled 0 can not lead to another whose label is also 0, but a vertex not labelled 0 must lead to one whose label is 0, otherwise the former would have another label. As we have seen, a winning strategy consists of moving from any position to a position in the kernel.

A Curiosity In the graph games described, one of the players will win. A draw is not possible. Which player wins depends on whether the starting point has label 0 (the second player can win) or another label (the first player can win). Now a game can be such that its graph is too large to be drawn (think of chess, for instance). However, we shall now show that, without drawing the graph, games exist where we can determine which of the two players has a winning strategy, even if we can not hope to find out what that strategy is. Such a game is that invented by D. Gale, and called by him **GNIM.** Let n times m counters be laid out in m rows and n columns. A move consists of removing some counter, together with all counters above or to the right of it, and the whole rectangle thus defined. The player who takes the counter in the bottom row and leftmost column loses. In this game, one of the players must eventually lose, and we prove that it must be the second player-provided, of course, that the first uses a winning strategy.

Assume, to the contrary, that the second player has a winning strategy. Let the first player start by removing the top counter in the rightmost column. Then the second player makes a move of his winning strategy. But this produces a configuration which the first player could have produced to begin with, as a move of the winning strategy. Since it is impossible that both players should have a winning strategy the assumption that the second player has a winning strategy is disproved. It is the first player who wins, but how we do not know.

Reference

1. Bouton, C.L.(1902) Nim, a game with a complete mathematical theory. Annals of Math.,3, 35-39.

4

Mathematical Programming

The studies in this chapter treat topics of historical and technical interest. 4.1 explains the role of Farkas' work as a foundation for duality theory. 4.2 is an entertainment for racing enthusiasts. 4.3 shows how an optimizing problem expressed in terms of probability can be transformed into conventional mathematical programming form; the text contains also some tables of a distribution that the reader might be hard put to find readily elsewhere. 4.4 is, again, historical in emphasis, comparing the simplex method with other approaches. 4.5 glances at dynamic programming, and 4.6 resolves an interesting paradox. 4.7 discusses duality in a wider context than usual and provides a surprising cue for geometrical issues treated more specifically in chapter 8.1.

4.1 THEOREMS OF ALTERNATIVES: A prehistory of linear programming

In 1894 Julius Farkas wrote a paper in Hungarian, in which he proved the following theorem:
Let

$$\begin{aligned} &a_{11}x_1+\ldots+a_{1n}x_n\geq 0 \\ &\ldots\ldots\ldots\ldots\ldots \\ &a_{m1}x_1+\ldots+a_{mn}x_n\geq 0 \end{aligned} \qquad (4.1.1)$$

be a set of homogeneous, linear inequalities, and let every solution $X_n=(x_1,\ldots,x_n)$ also satisfy the homogeneous inequality

$$b_1x_1+\ldots+b_nx_n\geq 0. \qquad (4.1.2)$$

Then there exists non-negative $Y_m=(y_1,\ldots,y_m)$ such that

$$\begin{aligned} &a_{11}y_1+\ldots+a_{m1}y_1=b_1 \\ &\ldots\ldots\ldots\ldots\ldots \\ &a_{1n}y_1+\ldots+a_{mn}y_m=b_n. \end{aligned} \qquad (4.1.3)$$

This theorem, often called the Farkas Lemma, became more widely known through a paper in German [1], which is its usual source. There is a trivial inverse theorem:

If there exists non-negative Y_m such that (4.1.3) holds, then any solution of (4.1.1) is also a solution of (4.1.2).

The Farkas Lemma can equivalently be expressed as follows:

Either (4.1.3) has a solution Y_m, or (4.1.1) and

$$b_1x_1+\ldots+b_nx_n<0$$

have a solution X_n.

Note that here, and throughout, we use repeatedly the notation X_n, Y_m for the row vectors $(x_1,\ldots,x_n)$, $(y_1,\ldots,y_m)$.
The two systems can not have solutions simultaneously, because if

$$\bar{X}_n=(\bar{x}_1,\ldots,\bar{x}_n) \textit{ and } \bar{Y}_m=(\bar{y}_1,\ldots\bar{y}_m)$$

were such solutions, then also

$$\bar{x}_1(a_{11}\bar{y}_1+\ldots+a_{m1}\bar{y}_m)+\ldots+\bar{x}_n(a_{1n}\bar{y}_1+\ldots+a_{mn}\bar{y}_m)\geq 0$$
$$\textit{and}$$
$$\bar{y}_1(a_{11}\bar{x}_1+\ldots+a_{1n}\bar{x}_n)+\ldots+\bar{y}_m(a_{m1}\bar{x}_1+\ldots+a_{mn}\bar{x}_n)<0,$$

which is contradictory.

Farkas was a physicist in Kolozsvár (now Cluj) in Transylvania, interested in problems of analytical mechanics, which had occupied, for instance, Fourier and Gauss before him. From our point of view the importance of the Farkas Lemma resides in the fact that it was the seed for the development of algebraic linear optimisation theory, leading to the Duality Theorem of Linear Programming. We give a proof of the Lemma (though not that of Farkas) in the Notes. Farkas observed that if in (4.1.1) some of the inequalities are replaced by equations, then such an equation can be replaced by a pair of inequalities. This has the effect that the corresponding y_j is replaced by $y_{j1}\geq 0$, and $y_{j2}\geq 0$, while in (4.1.3) y_j is replaced by $y_{j1}-y_{j2}=z_j$, say, which is not sign-restricted. We have then

Theorem 1. Let the system

$$\begin{array}{r}a_{11}x_1+\ldots+a_{1n}x_n\geq 0\\ \ldots\ldots\ldots\ldots\\ a_{p1}x_1+\ldots+a_{pn}x_n\geq 0\\ a_{p+1\,1}x_1+\ldots+a_{p+1\,n}x_n=0\\ \ldots\ldots\ldots\ldots\\ a_{q1}x_1+\ldots+a_{qn}x_n=0\end{array}\qquad (4.1.1')$$

be given, and let a solution $X_n=(x_1,\ldots,x_n)$ also satisfy (4.1.2), then there exist non-negative $y_1,\ldots,y_p$, and $y_{p+1},\ldots,y_q$ without sign restriction such that

$$a_{11}y_1+\ldots+a_{p1}y_p+a_{p+1\ 1}y_{p+1}+\ldots+a_{q1}y_q=b_1$$
$$\ldots\ldots\ldots\ldots\ldots\ldots \qquad (4.1.3')$$
$$a_{1n}y_1+\ldots+a_{pn}y_p+a_{p+1\ n}y_{p+1}+\ldots+a_{qn}y_q=b_n.$$

We obtain a further modification of the Lemma by adjoining n more inequalities $x_1\geq 0,\ldots,x_n\geq 0$. Then, if all solutions of (4.1.1) and $\boldsymbol{X}_n\geq 0$ satisfy also (4.1.2), there exist non-negative $\boldsymbol{Y}_m$ and $\boldsymbol{Z}_n=(z_1,\ldots,z_n)$ such that

$$a_{11}y_1+\ldots a_{m1}y_m+z_1=b_1,\ \textit{that is}\quad a_{11}y_1+\ldots+a_{m1}y_m\leq b_1$$
$$\ldots\ldots\ldots\ldots\ldots\ldots\ldots\ldots$$
$$a_{1n}y_1+\ldots+a_{mn}y_m+z_n=b_n,\ \textit{that is}\quad a_{1n}y_1++a_{mn}y_m\leq b_n.$$

In the same way as we derived Theorem 1 from the Lemma we obtain

Theorem 2. Let the system (4.1.1') be given, and let $\boldsymbol{X}_n\geq 0$ satisfy also (4.1.2), then there exist non-negative $y_1,\ldots,y_p$, and $y_{p+1},\ldots,y_q$ without sign restriction such that

$$a_{11}y_1+\ldots+a_{p1}y_p+a_{p+1\ 1}y_{p+1}+\ldots+a_{q1}y_q\leq b_1$$
$$\ldots\ldots\ldots\ldots\ldots\ldots\ldots\ldots$$
$$a_{1n}y_1+\ldots+a_{pn}y_p+a_{p+1\ n}y_{p+1}+\ldots+a_{qn}y_q\leq b_n.$$

Briefly: The adjoining of non-negative x_i has the effect of changing the equality signs in (4.1.3') into the inequality signs ≤ 0.

Theorems 1 and 2 above concern systems of homogeneous, linear equations and inequalities. We proceed to prove theorems about non-homogeneous systems, which might be considered generalisations of Farkas' Lemma.

Theorem 3. Let every solution of the system

$$a_{11}x_1+\ldots+a_{1n}x_n\geq d_1$$
$$\ldots\ldots\ldots\ldots\ldots\ldots \qquad (4.1.4)$$
$$a_{m1}x_1+\ldots+a_{mn}x_n\geq d_m$$

be also a solution of

$$b_1x_1+\ldots+b_nx_n\geq c. \qquad (4.1.5)$$

Then there exists non-negative $\boldsymbol{Y}_m=(y_1,\ldots,y_m)$ such that

$$a_{11}y_1+\ldots+a_{m1}y_m=b_1$$
$$\ldots\ldots\ldots\ldots$$
$$a_{1n}y_1++a_{mn}y_m=b_n \qquad (4.1.6)$$

and

$$d_1y_1+\ldots+d_my_m\geq c,$$

provided that the system (4.1.4) has a solution X_n,say. (An analogous proviso in the Farkas Lemma would be redundant, since the system (4.1.1) has the obvious solution $X_n=0$.)

The proviso will be used to prove an intermediate proposition from which Theorem 3 will follow by an application of the Farkas Lemma.

Assume that all solutions of (4.1.4) are also solutions of (4.1.5) and that the proviso holds. Then:
(a) All solutions of

$$a_{11}x_1+\ldots+a_{1n}x_n-d_1x_{n+1}\geq 0$$
$$\ldots\ldots\ldots\ldots$$
$$a_{m1}x_1+\ldots+a_{mn}x_n-d_mx_{n+1}\geq 0$$
$$x_{n+1}>0 \quad (\textit{note that } x_{n+1}\neq 0)$$

are also solutions of

$$b_1x_1+\ldots+b_nx_n\geq cx_{n+1},$$

since otherwise there would be a solution of (4.1.4)

$$\left(\frac{x_1}{x_{n+1}},\ldots,\frac{x_n}{x_{n+1}}\right)$$

which is not a solution of (4.1.5), in contradiction to the initial assumption.
(b) All solutions of

$$a_{11}x_1+\ldots+a_{1n}x_n\geq 0$$
$$\ldots\ldots\ldots\ldots$$
$$a_{m1}x_1+\ldots+a_{mn}x_n\geq 0$$

are also solutions of

$$b_1x_1+\ldots+b_nx_n\geq 0$$

since otherwise

$$(\bar{x}_1+\lambda x_1,\ldots,\bar{x}_n+\lambda x_n)$$

with sufficiently large λ would be a solution of (4.1.4), but not of (4.1.5), contradicting the initial assumption.

Both (a) and (b) follow from the proviso and differ only in the assumption about x_{n+1} (which in (b) is implicitly zero). Together they imply the following intermediate proposition:

Let all solutions of (4.1.4) be also solutions of (4.1.5). Then all solutions of

$$a_{11}x_1+\dots+a_{1n}x_n-d_1x_{n+1}\geq 0$$
$$\dots\dots\dots\dots$$
$$a_{m1}x_1+\dots+a_{mn}x_n-d_mx_{n+1}\geq 0$$
$$x_{n+1}\geq 0$$

are also solutions of

$$b_1x_1+\dots+b_nx_n\geq cx_{n+1}.$$

By applying the Lemma it follows that there exist non-negative values $y_1,\dots y_m, y_{m+1}$ such that

$$a_{11}y_1+\dots+a_{m1}y_m=b_1$$
$$\dots\dots\dots\dots$$
$$a_{1n}y_1+\dots+a_{mn}y_m=b_n$$
$$-d_1y_1-\dots-d_my_m+y_{m+1}=-c,$$

that is

$$d_1y_1+\dots+d_my_m\geq c,$$

and this is Theorem 3.

Just as we derived Theorem 2 by adjoining non-negative x_i's we obtain

Theorem 4. If all solutions of (4.1.4) with $\boldsymbol{X}_n\geq 0$ are also solutions of (4.1.5), then there exist non-negative $\boldsymbol{Y}_m$ such that

$$a_{11}y_1+\dots+a_{m1}y_m\leq b_1$$
$$\dots\dots\dots\dots$$
$$a_{1m}y_1+\dots+a_{mn}y_m\leq b_n$$
$$d_1y_1+\dots+d_my_m\geq c.$$

It is left to the reader to modify Theorems 3 and 4 for the case where some of the inequalities in (4.1.4) are changed into equations.

Examples

1. All solutions of

$$x_1-2x_2\geq -5$$

$$x_1+x_2\geq 1$$

$$-4x_1-x_2\geq -2$$

satisfy also

$$-x_1-x_2>+3.$$

See Figure 4.1.1.
Therefore non-negative y_j exist such that

$$y_1+y_2-4y_3=-1, \qquad -2y_1+y_2-y_3=-1$$

and

$$-5y_1+y_2-2y_3\geq -3.$$

For instance, $y_1=5/12$, $y_2=1/4$, $y_3=5/12$.

2. All solutions of

$$x_1-2x_2\geq -5, \quad x_1+x_2\geq 1, \quad -4x_1-x_2\geq -2, x_1\geq 0, x_2\geq 0$$

satisfy also

$$-x_1-x_2\geq -2.$$

(See Figure 4.1.2).
Therefore non-negative y_j exist such that

$$y_1+y_2\leq -1, \quad -2y_1+y_2-y_3\leq -1, \textit{ and } -5y_1+y_2-2y_3\geq -2.$$

For instance, $y_1=y_2=0$, $y_3=1$.

Figure 4.1.1

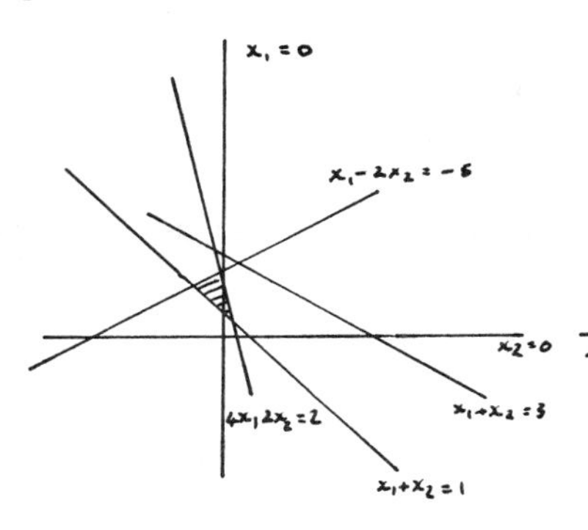

Figure 4.1.2

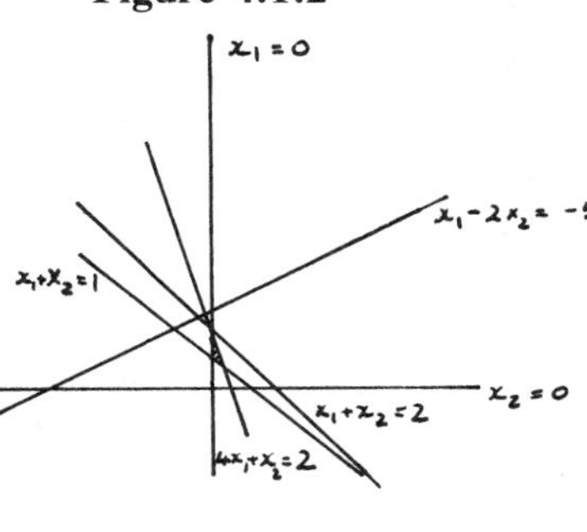

Notes

Let the matrix $\boldsymbol{A}$ be given. Call S the set of vectors $\boldsymbol{v}$ for which a non-negative vector $\boldsymbol{y}$ exists such that

$$\boldsymbol{A}^{\mathrm{T}}\boldsymbol{y}=\boldsymbol{v}.$$

This set is bounded and not empty (it contains $\boldsymbol{0}$, for which $\boldsymbol{y}=\boldsymbol{0}$).

We formulate the Farkas Lemma as follows:
Assume that the vector $\boldsymbol{b}$ is not in S, i.e. that no vector $\boldsymbol{y}\geq\boldsymbol{0}$ exists for which $\mathbf{A}^{\mathrm{T}}\mathbf{y}=\mathbf{b}$. Then there exists a vector $\mathbf{x}$ such that
(I) $\boldsymbol{Ax}\geq\boldsymbol{0}$, and (II) $\boldsymbol{b}^{\mathrm{T}}\boldsymbol{x}<\boldsymbol{0}$.

Proof The set S is closed, i.e. it contains its boundary. Let the vector $\underline{\boldsymbol{u}}\in S$ be such that

$$\min(|\boldsymbol{u}\text{-}\underline{\boldsymbol{b}}|:\boldsymbol{u}\in S)=|\underline{\boldsymbol{u}}\text{-}\boldsymbol{b}|.$$

Since $\underline{\boldsymbol{u}}\in S$ and $\boldsymbol{b}\notin S$, $|\underline{\boldsymbol{u}}\text{-}\boldsymbol{b}|>\boldsymbol{0}$ (i.e.not $\boldsymbol{0}$ but larger).

We show that the vector $\underline{\boldsymbol{x}}=\underline{\boldsymbol{u}}\text{-}\underline{\boldsymbol{b}}$ has the properties (I) and (II). But first we prove the following two statements:

(i) When $\boldsymbol{u}\in S$, then $\underline{\boldsymbol{x}}^{\mathrm{T}}(\boldsymbol{u}\text{-}\underline{\boldsymbol{u}})\geq\boldsymbol{0}$, and
(ii) $\underline{\boldsymbol{u}^{\mathrm{T}}\boldsymbol{x}}=\boldsymbol{0}$.

Proof of (i)
Define $\boldsymbol{u}(t)=t\boldsymbol{u}+(1\text{-}t)\underline{\boldsymbol{u}}$. Let $b\text{-}\boldsymbol{u}(t)$ have its minimum at $t=t_0$. This t_0 turns out to be

$$t_0=\frac{-(b-\underline{u})^T(\underline{u}-u)}{(\underline{u}-u)^2}.$$

If $t_0=0$, then $\boldsymbol{u}(t_0)=\underline{\boldsymbol{u}}$. If $t_0<0$, then $\boldsymbol{u}(t_0)$ is outside S: if $t_0>0$, then $\boldsymbol{u}(t_0)$ is inside. But this last case is impossible because, if so, $|\boldsymbol{u}(t_0)\text{-}\boldsymbol{b}|$, though inside S, would be nearer to $\boldsymbol{b}$ than $\underline{\boldsymbol{u}}$, which contradicts the definition of $\underline{\boldsymbol{u}}$. Thus $t_0=0$, or $t_0<0$. In either case

$$\underline{x}^T(u-\underline{u})=(\underline{u}-b)^T(u-\underline{u})=-t_0\geq 0.$$

This proves (i).

Proof of (ii)
Assume, by way of contradicting the statement, that $\underline{\boldsymbol{u}}^{\mathrm{T}}\underline{\boldsymbol{x}}$ is positive, or negative, but not zero. In the former case let $\boldsymbol{u}=½\underline{\boldsymbol{u}}\in S$. Then $\underline{\boldsymbol{x}}^{\mathrm{T}}(\boldsymbol{u}\text{-}\underline{\boldsymbol{u}})=½(\text{-}\underline{\boldsymbol{x}}^{\mathrm{T}}\underline{\boldsymbol{u}})<0$, contradicting (i). In the latter case, let $\boldsymbol{u}=2\underline{\boldsymbol{u}}\in S$. Then $\underline{\boldsymbol{x}}^{\mathrm{T}}(\boldsymbol{u}\text{-}\underline{\boldsymbol{u}})=\underline{\boldsymbol{x}}^{\mathrm{T}}\boldsymbol{u}<0$, again contradicting (i). It follows that $\underline{\boldsymbol{x}}^{\mathrm{T}}\underline{\boldsymbol{u}}=0$ must hold. This proves (ii).

Since all rows of the matrix $\boldsymbol{A}$ are vectors in S ($\boldsymbol{y}$'s are then the unit vectors), $\boldsymbol{A}\underline{\boldsymbol{x}}\geq\boldsymbol{0}$, which is (I).

Also, from (ii),

$$b^T\underline{x} = (\underline{u}^T - \underline{x}^T)\underline{x} = -\underline{x}^T\underline{x}.$$

Since $\underline{x} \neq 0$,

$$b^T\underline{x} = -\underline{x}^T\underline{x} < 0.$$

References

1.Farkas,J.(1901) Theorie der einfachen Ungleichungen. J.f.d. reine und angewandte Mathematik, 124,1-27.
2. Prékopa, A. (1980) On the development of optimization theory. Amer. Math. Monthly, 87, 527-542.
3.Dantzig,G.B. (1991) Linear Programming. In *History of Mathematical Programming*. Elsevier,19-31.
4. Kuhn, H.W. (1991) Nonlinear Programming:a historical Note.Ibid.,82-96.

4.2 SAFE GAMBLING

The other day, my friend and I went to the races in Wolverhampton. There we noticed a board with the message:

1	Guiburn's Nephew	13:8
2	Well Wrapped	5:4
3	Astra Radieux	20:1
4	Member's Revenge	5:2

4 runners

My friend was rather interested in the odds 20:1. He understood this to mean that if he staked an amount on Astra Radieux he would win twenty times that amount if the horse won, and I told him that he would then even get his stake back. I warned him (he is rather naïve) that he would lose his whole stake if another horse won.

"Well", he said, "what about betting on all the horses? One of them must win." I agreed that was a reasonable idea, but how should the total amount he was prepared to stake be distributed over the runners? The answer must have something to do with the odds. After some thought he suggested that the distribution should guarantee the largest possible gain.

"What do you mean by *guarantee*?"

"I want to be sure of getting the amount below which my pay-off can not drop."

"Oh yes, what we mathematicians call a *maximin* strategy. But do you realise that the guaranteed amount might be less than your total stake?"

"I shall have to take that chance and, in any case, I shall have had some fun. Moreover I might win more than the guaranteed minimum. Who knows? Certainly not you cautious mathematicians."

I could have told him something about that, but I did not want to dampen his enthusiasm. So I merely asked:

"Do you know how to find the magical distribution?"

"No, I don't", he replied. "Do you?"

As a matter of fact I did. This was my argument.

Denote the sum you bet on horse H_i by x_i. Let the total you stake be unity, so that

$$\sum_i x_i = 1.$$

Denote by v the smallest amount you could win, whichever horse is first, once you have chosen your x_i. This is the amount you wish to maximise. Let the odds offered on horse H_i be a_i. Then, if H_i wins, you receive $x_i a_i$ and your stake. Thus, the bookmaker will pay you $(a_i+1)x_i = x_i c_i$, say. The smallest of these amounts is to be v, so

$$x_i c_i \geq v \qquad (i=1,\ldots,n)$$

which we write

$$x_i c_i - y_i = v, \qquad (y_i \geq 0)$$

and we have also $x_i \geq 0$.

Our problem is therefore this:

$$\begin{gathered}
\textit{maximise } v \\
\textit{subject to} \\
x_1 c_1 - y_1 = v \\
x_2 c_2 - y_2 = v \\
\cdots\cdots \\
x_n c_n - y_n = v \\
x_i \geq 0 \quad (i=1,\ldots,n) \\
y_i \geq 0 \quad (i=1,\ldots,n) \\
x_1 + x_2 + \ldots + x_n = 1.
\end{gathered}$$

My friend protested:

"You would twist any problem into one of Linear Programming because you would then know how to solve it. Well, on with the Simplex Method!"

"Yes", I replied,"it is indeed an LP problem, but a very simple one: we don't have to apply sophisticated methods."

This is what I showed my friend. We have

$$x_i = (v+y_i)/c_i. \quad (c_i = a_i + 1 \text{ is certainly not zero.})$$

Then

$$x_1 + \ldots + x_n = \Sigma_i (v+y_i)/c_i = 1,$$

that is,

$$v\Sigma_i(1/c_i) = 1 - \Sigma_i(y_i/c_i).$$

v will be largest if all y_i are zero. Then

$$x_i = (1/c_i)/\Sigma_i(1/c_i) \qquad \text{(the stakes)}$$

and

$$v = 1/\Sigma_i(1/c_i) \qquad \text{(the guarantee)} .$$

The guarantee will be smaller than 1 if $\Sigma_i(1/c_i) > 1$.
Here are the calculations.

Horse_i	Odds a_i	$c_i = a_i + 1$	$1/c_i$	x_i	$x_i c_i$
1	13:8	2.625	0.381	0.329	0.86
2	5:4	2.25	0.444	0.383	0.86
3	20:1	21	0.048	0.041	0.86
4	5:2	3.5	0.286	0.247	0.86
			1.159	1.000	

My friend used this scheme. Well Wrapped won, and my friend lost 0.14 times his total stake.

It would have been, in any case, unrealistic to assume that a bookmaker would offer odds that would leave him out of pocket after the race. But we have assumed that he obligingly accepted any fractional cash.

But what I could, or, perhaps, should, have told my friend when he thought that he might, by good luck, win more than the guaranteed minimum I had worked out for him, was this:- If he uses the maximin strategy, as explained , then there is no chance of his obtaining more than the guaranteed sum. When the y_i are zero, all $x_i c_i$ are equal to v; the punter receives the same amount whichever horse wins!

For those who are interested, here are odds offered at the notorious 1993 Grand National Steeplechase (which was declared void because of false starts):

Odds	No.of horses	Odds	No.of horses	Odds	No.of horses	Odds	No.of horses
13:2	1	14:1	2	40:1	5	150:1	2
8:1	1	20:1	2	50:1	2	200:1	2
9:1	2	25:1	5	66:1	8	250:1	1
10:1	1	33:1	1	100:1	4	500:1	1

In that case $\sum(1/c_i) = 4/3$. If a horse with odds 200:1 had won, and you had staked $(1/201)/(4/3) = 3/804 = 1/268$ on it, you would have won 600/804, nearly 3/4, and thus lost altogether nearly 1/4, in accordance with the theory.

4.3 PROBABILISTIC PROGRAMMING

Linear programming is a method of maximising or minimising a linear function whose variables are subject to linear constraints. Let us assume that the coefficients, and the values of the constraints, are random variables with known probability distributions. We might then require that the constraints hold at least with some given probability: thus

$$Prob(\sum_i a_i x_i \geq b) \geq \alpha. \qquad (4.3.1)$$

To cope with this, our aim is to try to replace such probabilistic constraints with equivalent deterministic ones. We
shall distinguish two cases:

I. The a_i are known, and b is a random variable.

II. b is known and the a_i are random variables.

To begin with, take case I and assume that b has a known distribution function

$$F(B) = Prob(b \leq B).$$

Define b_α by $F(b_\alpha)=\alpha$ if the probability function is continuous or, if discrete, by the smallest value of B for which $F(B) \geq \alpha$. Let $B = \Sigma_i a_i x_i$, then

$$Prob(\sum_i a_i x_i \geq b) \geq \alpha \qquad (4.3.2)$$

if and only if

$$\sum_i a_i x_i \geq b_\alpha. \qquad (4.3.3)$$

The two inequalities (4.3.2) and (4.3.3) are equivalent.

Examples

b is normally distributed with zero mean and unit variance.

1. $\alpha=0.5$, $b_\alpha=0$,
 $\text{Prob}(\Sigma_i a_i x_i \geq b) \geq 0.5$ is equivalent to $\Sigma_i a_i x_i \geq 0$.

2. $\alpha=0.75$, $b_\alpha=0.6745$,
 $\text{Prob}(\Sigma_i a_i x_i \geq b) \geq 0.75$ is equivalent to $\Sigma_i a_i x_i \geq 0.6745$.

3. b is discrete (0,1,2) and uniformly distributed, viz.
 $\text{Prob}(b=n) = ⅓\ (n=0,1,2)$. Then
 $\text{Prob}(\Sigma_i a_i x_i \geq b) \geq 0.5$ is equivalent to $\Sigma_i a_i x_i \geq 1$.

Turning now to Case II, consider the probabilistic constraint

$$\text{Prob}(\Sigma_i^n a_i x_i \geq b) \geq \alpha$$

where the a_i are random. Here we are concerned with the distribution of a sum of random variables. We can deal conveniently only with cases where such a distribution belongs to the same type as do the summands, that is where the distribution is *stable* (see Notes). We shall discuss the three stable distributions specially mentioned in the Notes.

Case II.1. a_i is normally distributed with mean m_i and variance s_i^2. Then $\Sigma_i a_i x_i$ is also normally distributed with mean $M = \Sigma_i m_i x_i$ and variance $S^2 = \Sigma_i s_i^2 x_i^2$. Thus

$$\frac{\sum_i a_i x_i - M}{S}$$

has the standard normal distribution with zero mean and unit variance. Therefore, if we find τ_N such that

$$Prob(\sum_i \frac{a_i x_i - M}{S} \geq \tau_N) = \alpha,$$

that is

$$\frac{1}{\sqrt{2\pi}} \int_{-\infty}^{\tau_N} e^{-\frac{1}{2}x^2} dx = 1 - \alpha,$$

then the probabilistic constraint

$$Prob(\sum_i a_i x_i \geq b) \geq \alpha$$

is equivalent to the deterministic constraint

$$M + \tau_N S \geq b.$$

Examples

1. $m_i = 0$, $s_i = 1$ (i=1,2,3)

$\alpha = 0.75$, $\tau_N = -0.6745$

Equivalent constraint $-0.6745 \surd(\Sigma_i x_i^2) \geq b$.

2. $m_1=0,\ m_2=1,\ m_3=-1,\ s_1=s_2=s_3=2$

$\alpha=0.75,\ \tau_N=-0.6745$

Equivalent constraint $x_2-x_3-1.349\sqrt{(\Sigma_i x_i^2)}\geq b$.

We deal similarly with the other two distributions explicitly mentioned.

Case II.2 a_i has a Cauchy distribution with parameters λ_i and μ_i, that is

$$f(a_i)=\frac{1}{\pi}\frac{\lambda_i}{\lambda_i^2+(a_i-\mu_i)^2}.$$

$\Sigma_i a_i x_i$ again has a Cauchy distribution with parameters $\Sigma_i\lambda_i x_i$ and $\Sigma_i\mu_i x_i$.

The deterministic equivalent to

$$Prob(\sum_i a_i x_i \geq b)\geq\alpha$$

is

$$\sum_i \mu_i x_i+\tau_C\sum_i \lambda_i x_i\geq b$$

where

$$\frac{1}{\pi}\int_{-\infty}^{\tau_C}\frac{dx}{1+x^2}=1-\alpha.$$

Example

$$\mu_1=0,\ \mu_2=1,\ \mu_3=-1,\ \lambda_1=\lambda_2=\lambda_3=2.$$

$$\alpha=0.75,\quad \tau_C=-1.$$

Equivalent constraint: $-2x_1-x_2-3x_3\geq b.$

Case II.3 a_i has a Lévy distribution with parameters p_i and q_i. Dropping subscripts this gives the probability density

$$f(a)=\sqrt{\frac{q}{2\pi}}\frac{\exp\{-\frac{q}{2(a-p)}\}}{(a-p)^{3/2}}, \quad (q>0,a>p).$$

The distribution, as well as being a member of the **stable** family, is closely related to the normal distribution. See the Notes for details. The equivalent constraint is

$$\sum_i p_i x_i + \tau_L (\sum_i \sqrt{q_i x_i})^2 \geq b, \textit{ where}$$

$$\frac{1}{\sqrt{2\pi}} \int_{-\infty}^{\tau_L} x^{-3/2} \exp(-\frac{1}{2x}) dx = 1-\alpha.$$

Example

$$p_1=0,\ p_2=1,\ p_3=-1,\ q_1=q_2=q_3=2$$
$$\alpha=0.75, \quad \tau_L=0.7557$$
Equivalent constraint
$$x_1-x_2+0.7557(\sqrt{2x_1}+\sqrt{2x_2}+\sqrt{2x_3})^2 \geq b.$$

It is apparent that if b as well as the a_i are random, belonging to the same stable family, then the argument for case II carries over by writing the constraint

$$\sum_{i=1}^{n} a_i x_i > +b \quad \textit{as} \quad \sum_{i=1}^{n+1} a_i x_i \geq 0,$$
with
$$a_{n+1}=-b, \textit{ and } x_{n+1}=1.$$

Notes: Let $F(x)$ be the distribution function of the random variable X, that is

$$F(x) = \text{Prob}(x \leq X)$$

with frequency, or density, function

$$f(x)=\mathrm{d}F(x)/\mathrm{d}x.$$

The function of the real variable t, which for many purposes can be regarded as a "dummy" variable, defined by

$$g(t) = \int_{range of x} e^{itx} f(x) dx \quad (i = \sqrt{-1})$$

is called the **characteristic function** of the distribution. When X is discrete the characteristic function can be regarded as a **generating function.** A useful rôle of both characteristic and generating functions is to yield properties of the distribution which may be hard to get from the distribution itself. For example, we can write

$$g(t) = E[e^{itX}]$$

so that

$$[\frac{d^n g(t)}{dt^n}]_{t=0} = i^n E[X^n],$$

which is useful for finding moments. Again, if

$$S = X_1 + \ldots + X_n,$$

and if each X is independent of the others,

$$E[e^{itS}] = \{E[itX]\}^n = g^n(t).$$

For instance, the normal density

$$f(x) = \frac{1}{\sigma\sqrt{2\pi}} \exp\{-\frac{(x-m)^2}{2s^2}\}, \quad (-\infty < x < \infty)$$

has the characteristic function

$$g(t) = e^{mit - \frac{1}{2}s^2 t^2}.$$

Let $X_1, \ldots, X_n$ be independent and identically distributed normal random variables with means m and variances s^2, then the sum

$$\Sigma_i a_i x_i$$

has characteristic function $\exp(Mit - ½S^2t^2)$, where

$$M = \sum_i a_i m_i, \quad S^2 = \sum_i a_i^2 s_i^2.$$

This is again the characteristic function of a normal distribution.

In particular, if $X_1,\ldots,X_n$ each are independently and normally distributed with $m=0$ and $s=1$ (the **standard** normal), then the sum

$$\frac{(x_1+\ldots+x_n)}{n^{1/2}}$$

also has the characteristic function of the standard normal and so is itself standard normally distributed. This introduces the topic of **stable** distributions.

When the sum of n independent, identically distributed random variables divided by $n^{1/\alpha}$ has the same distribution as the individual terms in the sum, then the distribution of such a variable is said to be **stable** with index α. Thus, the normal distribution is stable with index 2. Stable distributions are known to exist for $0\leq\alpha\leq2$ and only those with $\alpha=½,1$ and 2 can be expressed in closed form. The stable distribution with $\alpha=1$ is the Cauchy distribution with density

$$f(x)=\frac{\lambda}{\pi\{\lambda^2+(x-\mu)^2\}}, \qquad (-\infty<x<\infty)$$

and characteristic function

$$e^{\mu it-\lambda t}, \qquad t>0.$$

In particular, the standard Cauchy distribution has density $f(x)=\{\pi(1+x^2)\}^{-1}$ and characteristic function $\exp(-t)$, and so has the sum $(x_1+\ldots+x_n)/n$.

The stable distribution with index ½ is the Lévy distribution with density

$$f(x)=\frac{1}{q\sqrt{2\pi}}\left(\frac{x-p}{q}\right)^{-3/2}\exp\left(-\frac{q}{2(x-p}\right), \quad q>0, \quad x\geq p,$$

and has characteristic function

$$\exp(pit-(1+i)\sqrt{tq}), \quad t>0.$$

It is easily seen that the sum $(x_1+\ldots+x_n)/n^2$ of standardised Lévy variables ($p=0,q=1$) is also a standardised Lévy variable.

[1] is a reference to one of Lévy's papers. In [2], Feller discusses extensively stable distributions and their characteristic functions.

We feel it useful to close these Notes with tables of the standardised forms of the three stable distributions discussed. In the case of the Lévy distribution we give also percentage points. Computationally there is not much of interest for the table-maker. Many ways have been proposed for the **error function** form of the normal distribution and there is a very efficient series form based on Chebyshev fitting which can be found in [3]. The distribution function of the Cauchy distribution is expressible in terms of inverse tangents and both distribution function and percentage points present no problems. The Lévy distribution can be expressed in terms of the normal. Indeed, the general form is

$$F(x)=\sqrt{\frac{2}{\pi}}\int_{\sqrt{\frac{q}{x-p}}}^{\infty} e^{-\frac{u^2}{2}}du=2\{1-Erf(\sqrt{\frac{q}{x-p}})\}$$

where

$$Erf(x)=\frac{1}{\sqrt{2\pi}}\int_{-\infty}^{x} e^{-\frac{y^2}{2}}dy.$$

References

1.Lévy, P. (1939) Sur certains procéssus stochastiques homogènes. Compos. Math.,7,283-339. See also [2] below for more references.

2.Feller, W.(1966) *An introduction to probability theory and its applications*. Vol.II. Wiley.

3.Press, W.H. et al. (1986) *Numerical recipes*. CUP.

Table 4.3.1 Standardized Normal and Cauchy Distributions

x	Normal	Cauchy	x	Normal	Cauchy
-2	0.023	0.148	0.1	0.540	0.532
-1.9	.029	.154	0.2	.579	.563
-1.8	.036	.161	0.3	.618	.593
-1.7	.045	.169	0.4	.655	.621
-1.6	.055	.178	0.5	.691	.648
-1.5	.067	.187	0.6	.726	.672
-1.4	.081	.197	0.7	.758	.694
-1.3	.097	.209	0.8	.788	.715
-1.2	.115	.221	0.9	.816	.733
-1.1	.136	.235	1.0	.841	.750
-1	.159	.250	1.1	.864	.765
-0.9	.184	.269	1.2	.885	.779
-0.8	.212	.285	1.3	.903	.791
-0.7	.242	.306	1.4	.919	.803
-0.6	.274	.328	1.5	.933	.813
-0.5	.309	.352	1.6	.945	.822
-0.4	.345	.379	1.7	.955	.831
-0.3	.382	.407	1.8	.964	.839
-0.2	.421	.437	1.9	.971	.846
-0.1	.460	.468	2.0	0.977	.852
0	0.500	0.500			

Table 4.3.2 Standardised Lévy Distribution

Percentage	Points	x	F(x)
0.01	0.1507	0	0
0.02	0.1848	0.1	0.0016
0.03	0.2123	0.2	0.0254
0.04	0.2371	0.3	0.0679
0.05	0.2603	0.4	0.1138
0.06	0.2827	0.5	0.1573
0.07	0.3046	0.6	0.1967
0.08	0.3263	0.7	0.2320
0.09	0.3479	0.8	0.2636
0.10	0.3696	0.9	0.2918
0.15	0.4826	1.0	0.3173
0.20	0.6089	1.1	0.3404
0.25	0.7557	1.2	0.3613
0.30	0.9309	1.3	0.3805
0.35	1.1449	1.4	0.3980
0.40	1.4118	1.5	0.4142
0.45	1.7524	1.6	0.4292
0.50	2.1981	1.7	0.4431
0.55	2.7986	1.8	0.4561
0.60	3.6364	1.9	0.4682
0.65	4.8567	2.0	0.4795
0.70	6.7353	2.1	0.4902
0.75	9.8492	2.2	0.5002
0.80	15.5800	2.3	0.5096
0.85	27.9596	2.4	0.5186
0.90	63.3279	2.5	0.5271
0.95	254.312	2.6	0.5351
		2.7	0.5428

4.4 CONTEMPORARIES AND LAUREATES

4.4.1 It has happened before that an idea, the solution of an old problem, some method, turned up in the mind of more than one individual independently and almost simultaneously. The most celebrated such case is the birth of calculus, the brainchild of Leibniz as well as of Newton. In the present essay we deal with a more modest topic, that of Linear Programming: the problem of maximizing a linear function subject to linear constraints. A satisfactory method for solving such problems was devised by George B. Dantzig: the **Simplex Method** [1]. In what follows we shall assume that the reader is familiar with its detail. It was developed in the 1940's and applied to problems of production and allocation in micro-economic situations. At about the same time Leonid Vital'evich Kantorovich developed **his** method for dealing with economic resources in the planned economy of the Soviet Union - macro-economic problems. We shall not give a complete description of Kantorovich's method, but rather compare the two methods by considering the following problem, fashioned, though somewhat simplified, after an example in [3].

It is required to manufacture n products by m production units (factories, machines...). If production unit i $(1,2,\ldots,m)$ produces product j $(1,2,\ldots,n)$ then it produces a_{ij} units in unit time. The fraction of working time of production unit i on product j is denoted by x_{ij}, and it is required that

$$\sum_j x_{ij}=1 \qquad (all\ i)$$

$$\sum_i a_{ij}x_{ij}=z \qquad (all\ j)$$

$$x_{ij}\geq 0.$$

The total output is to be maximized. The unknown variables are the x_{ij}. This will be recognized as a linear programming problem with z to be maximized.

Let

$$(a_{ij})=\begin{pmatrix}24 & 62.5 & 75\\ 12 & 0 & 150\\ 48 & 50 & 66\\ 0 & 30 & 54\end{pmatrix}.$$

Assume that we have reached, or started with, Plan I:

$$(x_{ij})=\begin{pmatrix} 1 & 0 & 0 \\ 1 & 0 & 0 \\ 0.19 & 0.81 & 0 \\ 0 & 0.16 & 0.84 \end{pmatrix}.$$

From the corresponding Simplex tableau we quote the relevant columns:

		150	
		x_{23}	
24	x_{11}	0	1
12	x_{21}	1	1
48	x_{31}	-0.82	0.19
50	x_{32}	0.82	0.81
30	x_{42}	-2.27	0.16
54	x_{43}	2.27	0.84
		-81.88	135.78

The 'shadow' cost in the x_{23} column is negative and therefore we make this variable positive. But we must avoid thereby making some other variable negative. Now the smallest positive ratio of the two values in the same row is

$$0.84/2.27=0.37$$

in the x_{43} row. This then is the variable to be made zero in exchange for x_{23}. This produces Plan II:

$$(x_{ij})=\begin{pmatrix} 1 & 0 & 0 \\ 0.63 & 0 & 0.37 \\ 0.49 & 0.51 & 0 \\ 0 & 1 & 0 \end{pmatrix}.$$

The relevant columns in the new Simplex tableau are now:

		62.5	
		x_{12}	
24	x_{11}	1	1
12	x_{21}	0.12	0.63
48	x_{31}	-0.90	0.49
50	x_{32}	0.90	0.51
30	x_{42}	0	1
159	x_{23}	-0.12	0.37
		-53.26	166.08

The shadow cost is again negative. We exchange x_{12} for x_{32} and notice that

$$0.51/0.90 = 0.56.$$

This gives Plan III.

$$(x_{ij}) = \begin{pmatrix} 0.44 & 0.56 & 0 \\ 0.56 & 0 & 0.44 \\ 1 & 0 & 0 \\ 0 & 1 & 0 \end{pmatrix},$$

and the relevant extract from the Simplex tableau is

		50	
		x_{32}	
24	x_{11}	-0.12	0.44
12	x_{21}	-0.13	0.56
48	x_{31}	1.12	1
62.5	x_{12}	1.12	0.56
30	x_{42}	0	1
150	x_{23}	0.14	0.44
		90.32	196.28

Here we stop. We shall see later that we have constructed the optimal plan.

We turn now to the procedure of Kantorovich. To begin with we compute *resolving multipliers* c_i and *objectively determined valuations* d_j by $c_i = a_{ij}d_j$ for those pairs of subscripts (i,j) for which x_{ij} is positive in the plan under consideration. Thus, for Plan I

$$c_1 = 24d_1,\ c_2 = 12d_1,\ c_3 = 48d_1,\ c_3 = 50d_2,\ c_4 = 30d_2,\ c_4 = 54d_3.$$

These equations are homogeneous so we can put $d_1 = 1$ and then:

$$c_1 = 24,\ c_2 = 12,\ c_3 = 48,\ c_4 = 28.8,\ d_1 = 1,\ d_2 = 0.96,\ d_3 = 0.533.$$

Also, we complete a table of $c_i = a_{ij}d_j$ for $i = 1,2,3,4$ and $j = 1,2,3$:

$i\backslash j$	1	2	3	t_i
1	24	60	40	60
2	12	0	80	80
3	48	48	35.2	48
4	0	28.8	28.8	28.8

In the final column we have quoted the largest amount in the corresponding column. These entries are of importance, in view of the following

Theorem A plan is optimal if $x_{ij}=0$ whenever $a_{ij}d_j<t_i$.

Proof

$$(\sum_j d_j)z=\sum_j\sum_i a_{ij}x_{ij}=\sum_i\sum_j (d_ja_{ij})x_{ij}$$

and this is

$$\sum_i\sum_j t_ix_{ij}=\sum_i t_i,$$

while, if the x_{ij} are not such that x_{ij} is always 0 when $a_{ij}d_j<t_i$, then we obtain $(\sum_j d_j)z \leq \sum_i t_i$. Now, if t_i does not appear where $x_{ij}\geq 0$, then it must appear where $x_{ij}=0$. This is, then, a sign that that particular x_{ij} ought to be increased. In Plan II, t_2 is at $j=3$, while $x_{23}=0$. Its increase must, as in the Simplex method be such that no other variable must thereby become negative. In the same way as in that method this leads to the exchange of the variables x_{23} and x_{43}, that is to Plan II.

For those (i,j) for which x_{ij} is now positive we compute again
$c_i = a_{ij}d_j$ from

$c_1 = 24d_1$, $c_2 = 12d_1$, $c_3=48d_1$, $c_3 = 50d_2$, $c_2 =150d_3$, $c_4 = 30d_2$,

that is,
$c_1=24$, $c_2=12$, $c_3=48$, $c_4=28.8$, $d_1=1$, $d_2=0.96$, $d_3=0.08$,

and the completed table reads:

$i\backslash j$	1	2	3	t_i
1	24	60	6	60
2	12	0	12	12
3	48	48	5	48
4	0	28.8	4	28.8

$t_1=60$ at (1,2), but $x_{12}=0$, so we increase it and obtain, just as with the Simplex method, Plan III.

Now we find

$c_1=24d_1,\ c_1=62.5d_2,\ c_2=12d_1,\ c_2=150d_3,\ c_3=48d_1,\ c_4=30d_2,$

that is,

$c_1=24,\ c_2=12,\ c_3=48,\ c_4=11.5,\ d_1=1,\ d_2=0.4,\ d_3=0.08.$

The completed table is now:

$i\backslash j$	1	2	3	t_i
1	24	24	6	24
2	12	0	12	12
3	48	20	5	48
4	0	12	4	12

The maxima are all at places where $x_{ij}\neq 0$, and it follows from the Theorem that we have with Plan III reached the optimum. The relationship with the Duality Theorem of Linear Prigramming is elucidated in [8] as follows: Write the dual problem

$$\textit{Minimize} \sum_i \gamma_i \ \textit{subject to} \ \gamma_i+\alpha_{ij}\delta_i\geq 0.$$

Let

$$\gamma_i=c_i/\sum_j d_j,\ \delta_j=-d_j/\sum_j d_j.$$

At optimality (by "complementary slackness")

$$x_{ij}(\gamma_i+a_{ij}\delta_j)=0.$$

Hence, if $x_{ij}>0$, then $\gamma_i=-a_{ij}\ \delta_j=a_{ij}\ d_j\ /\Sigma d_j$, i.e. $c_i=a_{ij}\ d_j$, and if $x_{ij}=0$, then $\gamma_i\geq -a_{ij}\ \delta_j=a_{ij}\ d_j\ /\Sigma d_j$, i.e. max $c_i\geq a_{ij}d_j$. This is Kantorovitch's optimality criterion.

Kantorovich's paper [2] was introduced by T.C. Koopmans [4]. This was the main source for making the work known in the West. For critical assessment see also [5]. Koopmans and Kantorovich received jointly the Nobel Prize for Economics in 1975, and the Silver Medal of the British Operational Research Society was presented to N.I. Kantorovich on behalf of her late husband on 6 July, 1987, at a reception given by the Royal Society of London. Koopmans was the editor of [6], from which many of us learned for the first time about the Simplex Method.

4.4.2 George Joseph Stigler (1911-1991) was an economist with wide-ranging interests, who received the Nobel Prize for Economics in 1982. We are here concerned with his paper [7], "The Cost of Subsistence", in which he endeavours to find the minimum cost of obtaining the calories, protein, minerals and vitamins, adequate or optimum, for various income levels. His results are contained in a table 'which gives the composition and cost of the most economical diets, in August 1939, and 1944, for an active economist (weighing 70 kilograms) who lives in a large city'. We describe his findings relative to 1939. The paper includes a list of 77 commodities and their nutritional values per dollar of expenditure. A preliminary weeding reduces the list to 15 items. Thereafter the procedure is experimental, because there does not appear to be any direct method of finding the minimum of a linear function subject to linear conditions. Thus the goal of linear programming is succinctly stated. By making linear combinations of various commodities the list is further reduced to 9 entries. Below are quoted the nine commodities and their nutritive values, together with the assumed daily requirements of five nutrients which Stigler uses in his computations.

	Calories (1000)	Calcium (grams)	Vitamin A (1000 IU)	Riboflavin (mg)	Ascorbic Acid (mg)
Wheat Flour	44.7	2.0	-	33.3	-
Evap.Milk (Can)	8.4	15.1	26.0	23.5	60
Cheddar Cheese	7.4	16.4	28.1	10.3	-
Beef Liver	2.2	0.2	169.2	50.8	525
Cabbage	2.6	4.0	7.2	4.5	5369
Spinach	1.1	-	918.4	13.8	2755
Sweet Potatoes	9.6	2.7	290.7	5.4	1912
Lima Beans	17.4	3.7	5.1	38.2	-
Navy Beans	26.9	11.4	-	24.6	-
Requirement	3	0.8	5	2.7	75

Finally, the following combinations were arrived at:

	Quantity	Cost ($)
Wheat Flour	370 lb.	13.33
Evap. Milk	57 cans	3.84
Cabbage	111 lb.	4.11
Spinach	23 lb.	1.85
Navy Beans	285 lb.	16.80
Total Cost		39.93

It is of interest to compare this result with what linear programming prescribes.

We have to minimize

$$x_1 + \ldots + x_9$$

subject to

$$44.7x_1 + 8.4x_2 + \ldots + 26.9x_9 \geq 3$$
$$2.0x_1 + 15.1x_2 + \ldots + 11.4x_9 \geq 0.8$$
$$\cdots\cdots\cdots\cdots$$
$$60.0x_2 + \ldots\ldots + 1912x_7 \geq 75.$$

The detailed computations are given in [7]. Two iterations after a feasible solution without artificial variables has been obtained the Simplex tableau, without slack variables, looks as follows:

	x_3	x_4	x_7	x_8	
x_1	-0.9711	3.1368	-0.3604	1.7985	0.0355
x_9	1.9734	-6.5523	1.1366	-3.0638	0.0486
x_6	0.0384	0.0556	0.3343	-0.0601	0.0051
x_2	-0.2707	4.5396	-0.6825	2.3186	0.0086
x_5	-0.0167	0.0185	0.1922	0.0049	0.0112
	-0.2467	0.1982	-0.3797	-0.0019	0.1090

In the tableau, x_1=wheat flour, x_9=navy beans,x_6=spinach,x_2=milk, x_5=cabbage.

$0.1090 is the daily expense, quite close to Stigler's $39.93 per annum. In his result some ingredients were in fact oversupplied. However, this is not the final answer. One further Simplex iteration gives the optimum solution:

	x_3	x_2	x_7	x_8	
x_1	-0.7841	-0.6910	0.1112	0.1964	0.0295
x_9	1.5827	1.4434	0.1516	0.2827	0.0610
x_6	0.0417	-0.0122	0.3427	-0.0885	0.0050
x_4	-0.0596	0.2203	-0.1503	0.5107	0.0019
x_5	-0.0156	-0.0041	0.1950	-0.0045	0.0112
	-0.2349	-0.0437	-0.3499	-0.1031	0.1087

References

1. G.B. Dantzig. *Linear programming and extensions*. Princeton University Press. 1963.
2. L.V. Kantorovich. *Matematicheskie metody orginasatii plenirovaniia*. Leningrad University Press. 1939.
3. .. ..*The best use of economic resources*. (Transl. P.F. Knightsfield) Pergamon Press. 1965.
4. T.C. Koopmans. A note about Kantorovich's paper 'Mathematical methods of organising and planning production'.Management Science (1960) 6, 364-5.
5. A. Charnes and W.W. Cooper. On some works of Kantorovich, Koopmans and others. Management Science (1962) 8, 246-262.
6. T.C.Koopmans. *Activity analysis of production and allocation*. Wiley. 1951.
7. G.J. Stigler. The Cost of Subsistence. J. of Farm Economics. (1945), 27, 303-314.
8. Susan Powell (1995) Personal communication.

4.5 SEQUENTIAL DECISIONS

4.5.1 Factory output

We are given the use of a workshop with 180 machines for three days. One machine produces daily goods G_1 worth £36, or goods G_2 worth £27. Which goods should the machines produce during the three days so as to maximize the total value?

If this were all, all machines would obviously work all the time on the more profitable goods, viz. G_1. However, the machines are subject to breakdowns. The frequency of such misfortunes depends on the work they are doing. If machines work for a full day on G_1 only one third of their number will be operative during the next day. If they work on G_2 their availability the next day will be two thirds of their number. Table 1 gives more detail.

Table 4.5.1.1 Value of Possible Allocations

First Day	G_1 180 x 36 = 6480				G_2 180x27 =4860			
Second Day	G_1 60x36 =2160	G_2 60x27 =1620			G_1 120x36 =4320	G_2 120x27 =3240		
Third Day	G_1 20x36 =720	G_2 20x27 =540	G_1 40x36 =1440	G_2 40x27 =1080	G_1 40x36 =1440	G_2 40x27 =1080	G_1 80x36 =2880	G_2 80x27 =2160
Total	9360	9180	9540	9180	10620	9260	**10980**	10260

In principle, such a table could be constructed for any number of days, and for any number of goods, but it would soon become unmanageable for large numbers. We are therefore looking for a more economical method.

To formulate the problem algebraically, denote the number of machines producing goods G_1 on the i-th day by y_i, and the number of machines in working condition on the i-th day by z_i. We are given $z_1=180$. The value of goods produced on the first day will be $36y_1+27(180-y_1)$. On the second day, $z_2=\frac{1}{3}y_1 + \frac{2}{3}(180-y_1)$, and the value of goods produced will be $36y_2+27(z_2-y_2)$. On the third day, $z_3=\frac{1}{3}y_2+\frac{2}{3}(z_2-y_2)$, producing goods to the value of $36y_3 + 27(z_3-y_3)$. The y_i are to be chosen so that

$$36(y_1+y_2+y_3)+27(180+z_1+z_2-y_1-y_2-y_3)$$

is maximized. Moreover, all y_i and z_i must be non-negative. This is, of course, a linear programming problem, but we shall solve it by another method - that of **Dynamic Programming** [1]. Typically, Dynamic Programming works backwards. On the third day all z_3 available machines will naturally do the first job, $y_3=z_3$, producing goods worth

$$36y_3=12y_2 + 24(z_2-y_2).$$

Back to the second day! The total output during the last two days is worth

$$36y_2+27(z_2-y_2)+36y_3=51z_2-3y_2.$$

This is largest when $y_2=0$. It is then

$$51z_2=17y_1 +34(180-y_1)=6120-17y_1.$$

Now to the first day! The total output of the three days is

$$36y_1+27(180-y_1)+6120-17y_1=10980-8y_1.$$

This is largest if $y_1=0$.

Our conclusions are:

First day	produce G_2,	worth	4860
Second day	produce G_2,	worth	3240
Third day	produce G_1,	worth	2880
		Total	10980

If we had only two days to use the originally 180 machines, then all y_2 machines on the second day would have produced G_1, worth $4320-12y_1$, largest when $y_1=0$. On the first day, G_2 would be produced, worth 180x27=4860. The total output is worth 9180. This agrees, of course, with the information presented in Table 4.5.1.
[1] has been supplemented by possibly more attractive texts, of which [2] is recommended.

4.5.2 The Wrong Penny

We are given a collection of pennies, all except one genuine. We are told that the false one is heavier than the others, and we are challenged to find it, using only a balance and no weights in a minimum of weighings.

This is a severely simplified version of the *defective coin problem*. Extensive discussion and analysis will be found in [3], [4], [5] and [6].

To illustrate our strategy let us assume that we have seven pennies. Partition the set of 7 into 3 + 3 + 1. Put the two triples one on each arm of the balance.

If they balance, then all six are genuine and the false member remains. But if one of the triples is heavier we know that it contains the false coin. To identify which is the false coin, take two from the heavy triple and place one on each arm of the balance. If one is heavier, it is the false penny: otherwise it is the third unused coin.

Observe two key features of this procedure: we first partition into three portions, two of them equal, and we might finish in a single weighing, but can not be sure of it.

This example suggests what is to be done with an arbitrary number of coins, say n. Partition into $k+k+(n\text{-}2k)$. Weigh k against k. If they balance consider the remainder. If one of the k-tuples is heavier, examine it further. The first stage consists of just one weighing; the next might tackle k or $n\text{-}2k$ pennies. We want to find k so that the larger of k and $n\text{-}2k$ is as small as possible, given n. That smallest k, given n, we denote by k_n.

Now k may be $1,2,\ldots,[\frac{1}{2}n]$ (the largest integer not exceeding $\frac{1}{2}n$). On the other hand, n-$2k$ may be $(n-2)$, $(n-4),\ldots,(n-2[\frac{1}{2}n])$ (which may be 0 or 1). The first sequence is increasing and the second is decreasing, and they cross over as follows:

when $n \equiv 0 \pmod 3$ (i.e. when n is divisible without remainder by 3) at $n/3$, and this number appears in both sequences;

when $n \equiv 1 \pmod 3$ (i.e. when n is of the form $3m+1$) at $[n/3]+1$, which also appears in both sequences;

when $n \equiv 2 \pmod 3$ (i.e. n of form $3m+2$) at $[n/3]+1$, which appears in the second sequence.

All three values can be written $[(n+2)/3]$. They are values not smaller than the corresponding value in the other sequence, and are thus, because they appear at the crossing-over, equal to k_n. The number of weighings possibly necessary for given n will therefore be

$$f(n) \;=\; 1+f([(n+2)/3]).$$

Examples

$n=12$ $k=$ 1 2 3 4 5 6
n-$2k=$ 10 8 6 4 2 0
$k_n=\min(10,8,6,4,5,6)=4$

$n=13$ $k=$ 1 2 3 4 5 6
n-$2k=$ 11 9 7 5 3 1
$k_n=\min(11,9,7,5,5,6)=5$

$n=14$ $k=$ 1 2 3 4 5 6 7
n-$2k=$ 12 10 8 6 4 2 0
$k_n=\min(12,10,8,6,5,6,7)=5.$

We obtain the table:

n	1	2	3	4	5	6	7	8	9	10	11	12	13	14	15
k_n	1	1	1	2	2	2	3	3	3	4	4	4	5	5	5
$f(n)$	1	1	1	2	2	2	2	2	2	3	3	3	3	3	3

It is easily seen that

$$f(n)=m \quad (3^{m-1}+1 \leq n \leq 3^m).$$

Problem The reader might like to try the following version of the puzzle with 12 coins, all apparently identical, but one of which **may** be false in the sense of weighing differently from the others. Given a balance and no weights establish in not more than three weighings (a) whether one of the coins is false, (b) if so, which one it is and

(c) whether it is heavier or lighter than a genuine coin. The initial partition is in fact given by the table above. The reader has to provide the details.

References

1. Bellman, Richard (1957) *Dynamic programming*. Princeton University Press.
2. Smith, David, K. (1991) *Dynamic programming: a practical introduction.* Ellis Horwood. Chichester.
3. Smith, C.A.B. (1947) The counterfeit coin problem. Math. Gaz., 31,31.
4. O'Beirne, T.H. (1965) *Puzzles and paradoxes*. pp.33-48. OUP.
5. Steinhaus, H. (1950) *Mathematical snapshots*. pp.34-45. OUP.
6. Newman, D.J. Thoughtless Mathematics. The Mathematical Intelligencer. 15, 58-60.

4.6 COMMONSENSE OR PARADOX?

4.6.1 Mr. Walker intends to pack his rucksack with a selection from three items, A_1, A_2 and A_3. Their individual weights and space requirements are, respectively,

	A_1	A_2	A_3	
weight	3	1	2	lbs.
space	1	2	1	cu.ft.

The capacity of the rucksack is 4 cu. ft. He does not mind how much of any item he takes provided that the total weight does not exceed 9 lbs. and, of course, provided that the capacity is not exceeded. However, how much to take of each must depend on its utility. In this case the utility of an item is proportional to the amount carried and the items can be divided into fractions. The unit utilities, in some convenient measure, are

A_1	A_2	A_3
1	3	2

To put it in algebraic form, Mr. Walker has to solve the problem:

$$\textit{maximize}\ x_1+3x_2+2x_3=B$$

subject to

$$3x_1+x_2+2x_3=9$$

and a space limitation

$$x_1+2x_2+x_3=4$$

where x_i is the amount of A_i he packs and must, of course, be non-negative.

Mr. Walker knows, and we know, that this is a linear programming problem and that, because there are two constraints, not more than two of the variables will be positive at the maximum. Let us guess that the variable with value zero will be x_2. Then we have

$$x_1 = 1 + 3x_2,\ x_3 = 3 - 5x_2,\ B = 7 - 4x_2.$$

If x_2 were positive, B would be smaller. So, to take $x_2=0$ was a lucky guess, or an inspiration. B is then 7, $x_1=1$ and $x_3=3$.

Mr. Walker now cogitates: with a weight limitation of 9, I can obtain as much, but not more, than a total utility of 7. Presumably, to obtain 7 I must accept to have to carry a weight of 9. This is commonsense. Is it?

Let us turn the problem upside down. We want a total utility of 7. What is the least amount we have to carry to achieve this?

Formally we wish to

$$\textit{minimize}\ 3x_1+x_2+2x_3=C$$

$$\textit{subject to}$$

$$x_1+2x_2+x_3=4$$

$$x_1+3x_2+2x_3=7, \quad x_i \geq 0.$$

We again trust our knowledge of linear programming. This time, let $x_1=0$. Then

$$x_2=1-x_1, \quad x_3=2+x_1, \quad C=5+4x_1.$$

Thus, the minimum weight is obtained at $x_2=1$, $x_3=2$ and it is **only 5!** This is satisfactory, but what about commonsense? What has gone wrong?

To see the relationship between

$$x_1 + 3x_2 + 2x_3 = \text{U(tility)}$$

and

$$3x_1 + x_2 + 2x_3 = \text{W(eight)}$$

when also

$$x_1 + 2x_2 + x_3 = 4$$

we solve for the x_i. This is reasonable because we suspect that the apparent paradox is due to the fact that the x_i must be non-negative. After all, to solve linear equations is easy, but inequalities can play curious tricks.

We obtain

$$x_1 = 4 - \tfrac{3}{4}\text{U} + \tfrac{1}{4}\text{W} \geq 0$$
$$x_2 = 4 - \tfrac{1}{4}\text{U} - \tfrac{1}{4}\text{W} \geq 0$$
$$x_3 = -8 + 5\text{U}/4 + \tfrac{1}{4}\text{W} \geq 0.$$

Figure 4.6.1 Feasible Region

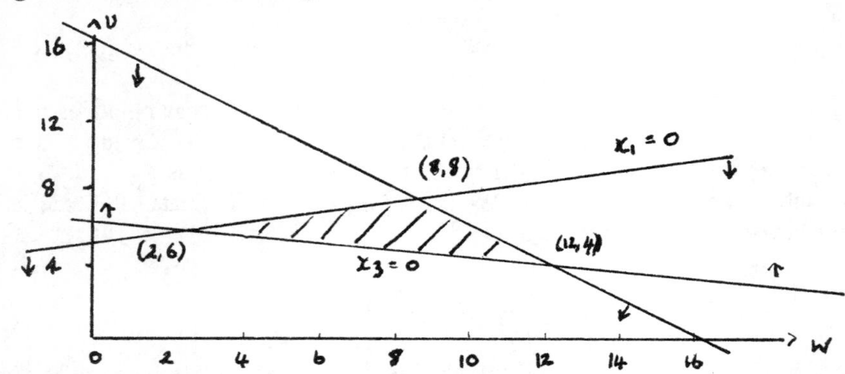

The feasible points lie inside the shaded triangle and on its boundary. For W=9 the maximum of U is indeed 7, but U can be 7 even when W=5. Observe that, had we started with $2 \leq W \leq 8$, the paradoxical result would not have appeared.

Because Figure 4.6.1 was so useful in explaining the situation, we spend some more time studying it. We discover a curious fact: we could obtain a higher utility than 7 for less than the 9 we were originally prepared to carry. For W=8 we need carry only 8!

To understand this we change our constraints from equations to inequalities:

$$3x_1 + x_2 + 2x_3 \leq 9$$
$$x_1 + 2x_2 + x_3 \leq 4.$$

We solve this system by the Simplex Method. Introducing the slack variables y_1 and y_2 we get the following tableaux:

	x_1	x_2	x_3			x_1	x_2	x_3	
y_1	2	1	2	9	y_1	1	-3	-2	1
y_2	1	2	1	4	x_3	1	2	1	4
	-1	-3	-2	0		1	1	2	8

The answer is Maximum 8, for $x_3=4$, $x_1=x_2=0$.

4.6.2 At the ports A and B, 6 and 1 units of some commodity are, respectively, available for shipping. At the ports C and D, respectively, 4 and 3 units are required. The cost of transportation is shown in the following table:

		To	C	D	available
From	A		10	3	6
	B		2	5	1
Required			4	3	units

It is easy to establish that the cheapest transport scheme is

		To	C	D	total
From	A		3	3	6
	B		1	0	1
Totals			4	3	

The total cost of this scheme is 41. The only snag is that the various port authorities have made a mistake. In fact, 4 units are required at D and, fortunately, 2 units are available at B. The cheapest transfer scheme is, in fact,

		To	C	D	total
From	A		2	4	6
	B		2	0	2

The total cost is 36.

But how can that be? We transport more, but pay less for it! Is this commonsense?

This is the **Doig Paradox**, after Alison Doig who first exhibited it to her friends a long time ago.

We might suspect that the paradoxical result is due to the fact that in the second case we could make better use of the relatively cheap transport from B to C because the restriction to 1 available unit was lifted. But did we take the most advantage of this relaxation? To investigate we write, in obvious notation,

$$x_{AC} + x_{AD} = 6 + y_1, \quad x_{AC} + x_{BD} = 1 + y_2,$$

$$x_{AC} + x_{BC} = 4 + y_3, \quad (x_{AD} + x_{BD} = 3 \text{ is implied}),$$

all x non-negative, and

$$\text{minimize } M = 10x_{AC} + 3x_{AD} + 2x_{BC} + 5x_{BD} .$$

The three equations above, i.e. the constraints, are equivalent to

$$x_{BC}=1+y_2-x_{BD}, \; x_{AC}=3-y_2+y_3+x_{BD}, \; x_{AD}=3+y_1+y_2-y_3-x_{BD},$$

and therefore

$$M = 41 + 3y_1 - 5y_2 + 7y_3 + 10x_{BD}.$$

As long as $y_1=y_2=y_3=x_{BD}=0$ the cost is 41. But M can be reduced by increasing y_2. Because of the second equation for x_{AC}, we can not increase y_2 to more than 3. Let us do this; then we can send

	To	C	D	Total
From	A	0	6	6
	B	4	0	4
Totals		4	6	

at a cost of $41\text{-}5y_2=26$. This is the best we can do under the circumstances.

On a similar matter, P. Rivett [1] looks at the problem

				Available
	(3)	(7)	(7)	3
	(4)	(9)	(6)	6
	(6)	(10)	(9)	5
Required	6	3	5	

(costs in parentheses)

Naively, one would use the cheapest route policy and distribute

3	0	0
3	3	0
0	0	5

at a cost of 93. But

0	0	3
4	0	2
2	3	0

costs only 82 although the most expensive route is fully used.

Reference

1. Rivett, P. (1968) *Concepts of operational research*. pp 37-38, Watts.

4.7 DUALITY AND OPTIMIZATION

4.7.1 The term *duality* appears in many branches of mathematics, for instance in topology, projective geometry, functional analysis, Boolean algebra, and others. We shall here consider a concept of duality defined as follows:

Consider a minimization problem with objective function $f(v)$, and a maximization problem with objective function $h(x)$, based on the same data, such that no feasible value of $h(x)$ exceeds any feasible value of $f(v)$. Then, if $f(v)=h(x)$, both $f(v)$ and $h(x)$ are optimal. Maximizing $h(x)$ and minimizing $f(v)$ are called *dual problems*.

Example 1:

$$\textit{Maximize } h(x)=\sum_{i=1}^{n} h_i x_i, \textit{ subject to}$$

$$\sum_{i=1}^{n} a_{ij}x_i + x_{n+j} = c_j \quad (j=1,..,m):$$

$$\textit{Minimize } f(v)=\sum_{j=1}^{m} c_j v_j, \textit{ subject to}$$

$$\sum_{j=1}^{m} a_{ij}v_j - v_{m+i} = h_i \quad (i=1,..,n)$$

$$\textit{with all non-negative} x_i, x_{n+j}, v_j, v_{m+i}.$$

In this case

$$f(v)-h(x)=$$

$$\sum_{j=1}^{m} v_j(c_j - \sum_{i=1}^{n} a_{ij}x_i) - \sum_{i=1}^{n} x_i(h_i - \sum_{j=1}^{m} a_{ij}v_j)$$

$$= \sum_{j=1}^{m} v_j x_{n+j} + \sum_{i=1}^{n} x_i v_{m+i} \geq 0.$$

The Duality Theorem of Linear Programming asserts that if one of the two problems has a finite solution, then so has the other, and that there exist x's and v's such that $h(x) = f(v)$. Hence these x and v produce the two extrema. For instance:

Minimize $20v_1+10v_2$, subject to $3v_1+v_2 \geq 3$, $4v_1+3v_2 \geq 6$:

Maximize $3x_1+6x_2$, subject to $3x_1+4x_2 \leq 20$, $x_1+3x_2 \leq 10$.

The constraints are satisfied by $v_1=0.6$, $v_2=1.2$, $x_1=4$, $x_2=2$. Then $f(v)=h(x)=24$, and 24 is the maximum of $h(x)$ as well as the minimum of $f(v)$.

If, however, one of the problems has no solution to its constraints, then the other may also be contradictory, or its extremum may be unbounded. For instance:
$h(x)=5x_1+x_2$, subject to $3x_1\text{-}2x_2 \leq 6$, $-4x_1+2x_2 \leq 4$, $(x_1,x_2 \geq 0)$.
We can write
$3x_1\text{-}2x_2+y_1=6$, $-4x_1+2x_2+y_2=4$, $(y_1,y_2 \geq 0)$,
that is,
$x_1=-10+y_1\text{-}y_2$, $x_2=-18+2y_1\text{-}1.5y_2$, $h(x)=-68+7y_1\text{-}6.5y_2$.
y_1 can be made large without limit, and so can $h(x)$, which has no finite maximum; but it is not contradictory: $x_1=x_2=1$ satisfy the constraints. As to the dual:
$f(v)=6v_1+4v_2$, $3v_1\text{-}4v_2 \geq 5$, $-2v_1+2v_2 \geq 1$, $(v_1,v_2 \geq 0)$.

The constraints are contradictory: multiply the second by 2 and add: then $v_1 \leq -7$ contradicts $v_1 \geq 0$.

An instance where both problems are contradictory is:

$h(x) = 2x_1 - x_2$, subject to $x_1 - x_2 \leq 1$, $-x_1 + x_2 \leq -2$, $(x_1, x_2 \geq 0)$

$f(v) = v_1 - 2v_2$, subject to $v_1 - v_2 \geq 2$, $-v_1 + v_2 \geq -1$, $(v_1, v_2 \geq 0)$.

Both sets of constraints imply that $0 \geq 1$.

4.7.2. Consider a network with two selected nodes, the source S and the terminal T, with a flow from S to T. At each node the total incoming flow equals the total outgoing flow, while the total flow starting at S equals the total flow terminating at T. Moreover, each link has a capacity which the flow through it may not exceed. Also consider cuts in the network, defined as sets of links such that, if they were omitted, no flow would be possible from S to T. Such a cut is, for instance, the set of all links emanating from S. It is obvious that no total flow can be larger than the total of all capacities of all the links in a cut. Hence, if we can exhibit such a cut and such a flow that their totals are equal, then we have found a maximal flow, and a minimal cut. It is, in fact, possible to prove the equality by the Duality Theorem of Linear Programming [1]. However, we shall show how to construct the maximal flow and the minimal cut, and it will emerge that they are equal. To illustrate the procedure we use the network of Figure 4.7.1, where the link capacities are given.

Figure 4.7.1 Capacities

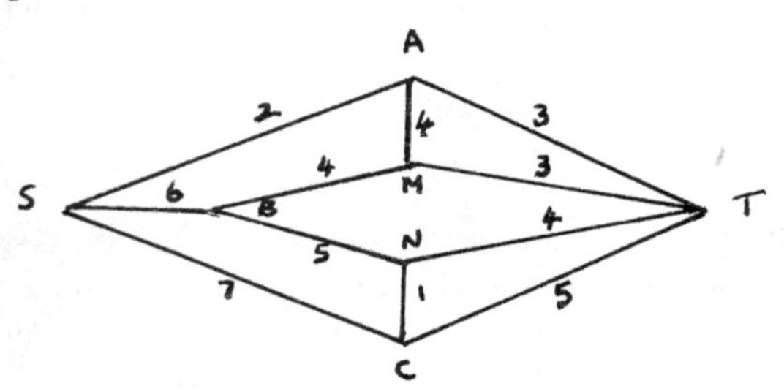

Start with some flow, for example zero flow if you can not think of a bigger one. Now see Figure 4.7.2.

Figure 4.7.2 First Flow

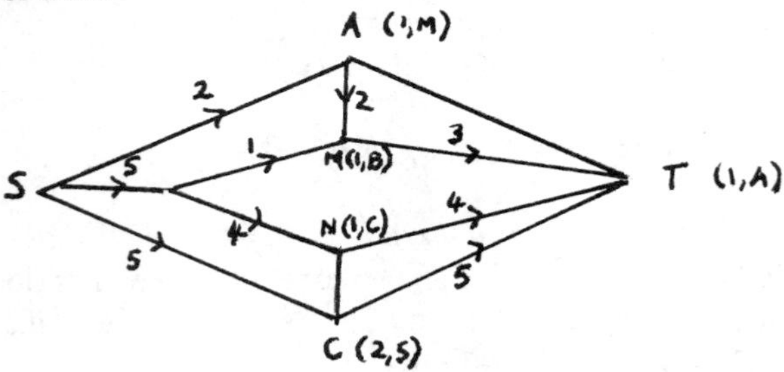

The total flow is 12. We increase a flow, if possible, by applying the following labelling rules:

1) S receives a label which is larger than the total capacity of the links starting at S. (It has no second label, but other nodes will have one.)

2) Take an unlabelled node V, say, at the end of a link starting at a labelled node, L, say (in the example, A, B or C).

V receives two labels: the first is the difference between the capacity of LV and its flow, but limited by the first label of L. This indicates a possible addition to the flow. The second label is 'L', where that additional flow could come from

3) If there is an unlabelled vertex, W, say, with a link to a labelled vertex, L, say, then the first label W receives the flow capacity of WL, but limited by the first label of L. This is a flow that could be withdrawn and redirected. In our example we have the labels:

B(1,S), C(2,S), M(1,B), N(1,C);
A(1,M);
T(1,A).

To label T means that the total flow can be increased by its first label. This leads to Figure 4.7.3.

Figure 4.7.3 **Second Flow**

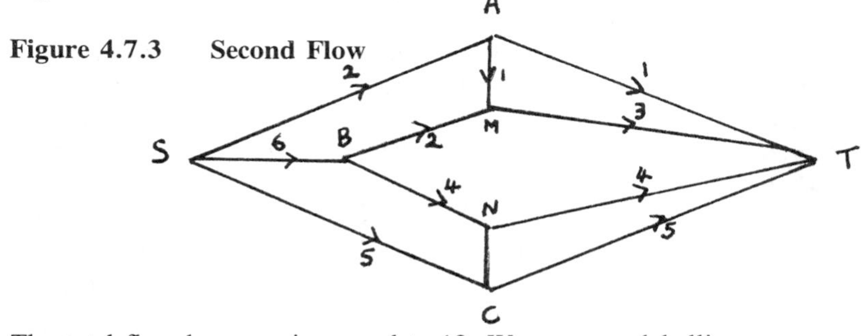

The total flow has now increased to 13. We carry on labelling:

C(2,S), N(1,C);
B(1,N);
M(1,B);
A(1,M), T(1,A).

Figure 4.7.4 **Third Flow**

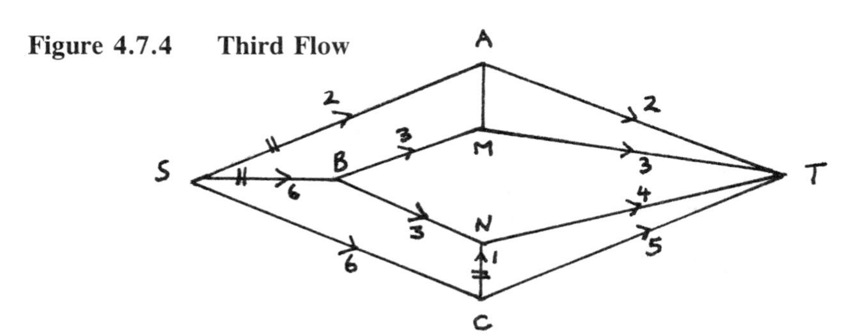

This leads to Figure 4.7.4, with a total flow of 14. Now the only possible further label is C(1,S), and T can not be labelled. We shall prove that this means that the maximum flow has been found, and we shall prove it by constructing a cut with the same total capacity as the flow.

Separate the nodes into two sets X and Y. X is the set of labelled nodes, Y that of unlabelled nodes. Neither set is empty since S is labelled and T is not. The cut we seek is the set of arcs from any vertex of X to any vertex of Y. In our example this is (SA,SB,CN,CT). This is a cut because it separates X from Y, hence S from T. There is no flow from Y into X, otherwise some member of Y could have been labelled. Hence all the flow is through links of the cut. These links are saturated, otherwise we could have continued labelling. Hence the total flow is equal to the total capacity of the cut. **QED**

In Figure 4.7.4 the cut has been indicated by two short lines through its links. The labelling procedure described is based on [2].

4.7.3. We call a graph, or a network, planar if it can be drawn on a plane without any links crossing. The networks in Section 4.7.2 are planar. We view the links of such a planar graph as boundaries of countries. We construct a dual, also planar, graph by putting a node into any country and connecting any two in adjacent countries. We attach to each new link a number equal to the capacity of the old link which it crosses.

Take now the network of Section 4.7.2 and draw a line from S to T outside the network. This produces two more countries, one with infinite area. Construct the dual network to the new one and consider a path between the nodes in these two new countries. Interpret the numbers attached to its links as lengths.

Any set of original links crossed by the path of the dual network constitute a cut in the original network, its capacity equal to the length of the path. Hence the shortest path in the new network is the path which crosses just the links of the minimal cut.

4.7.4. X_i $(i=1,2,3)$ are the vertices of a triangle with no angle exceeding 120°, and X is an internal point. Denote the sum of the straight line segments XX_i by f(X). Y_j $(j=1,2,3)$ are vertices of an equilateral triangle whose sides pass, respectively, through X_1, X_2 and X_3. Denote the altitude of this triangle by h. See Figure 4.7.5.

Figure 4.7.5

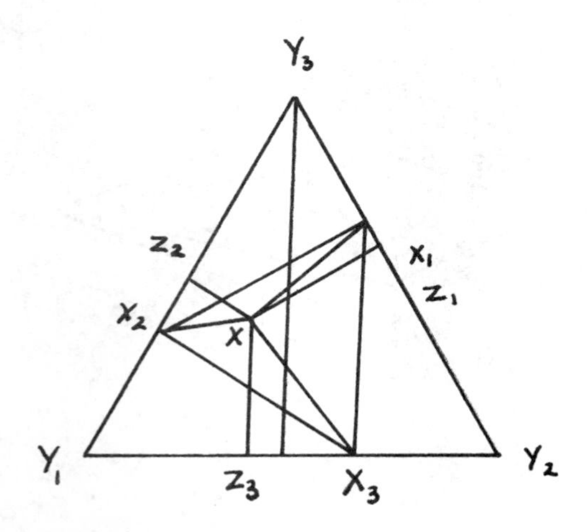

We shall show that $h \leq f(X)$. If a point X can be found and an equilateral triangle be drawn through the X_i such that $h=f(X)$, then X minimizes $f(X)$ and h maximizes the altitude of any of the equilateral triangles. In Figure 4.7.5 we have marked points Z_i as the feet of the perpendiculars from X on to the sides of the triangle $Y_1Y_2Y_3$. Now, in an equilateral triangle the sum of the segments XZ_i equals its altitude for any interior point X, and it is clear from the Figure that

$$XZ_1+XZ_2+XZ_3 \leq XX_1+XX_2+XX_3,$$

that is, $h \leq f(X)$.

$f(X)$ will be smallest for a point X such that $XX_1 \perp Y_2Y_3$,

$XX_2 \perp Y_1Y_3$, $XX_3 \perp Y_1Y_2$. In that case $f(X)=h$, and h will be the largest altitude of any equilateral triangle through the X_i. We call the point X the *Torricelli point*.

The minimizing problem was posed by P. Fermat early in the 17-th century, and the maximizing problem by Moss [3]. The precise history of these problems and of their solutions is somewhat obscure, but see [4] and [5]. Of course, none of this tells us how to find the Torricelli point. For this, see Chapter 8.1.

We are grateful to Professor J. Krarup of the University of Copenhagen for informative discussions on this topic.

4.7.5. It would, of course, be incorrect to assume that if $f(x)$ and $h(v)$ satisfy the condition in Section 4.7.1, then the two optima will be equal (as they are in ordinary LP). A counterexample is readily at hand. Take a linear programming problem in which at least one of the optima is not given in integer values of the variables, such as Example 1 in Section 4.7.1. If we now change the problem to one of "integer linear programming" with the further constraint that the values of the variables must be integers, then the maximum, if affected by this constraint, will be reduced, while the minimum, if affected, will be increased. In any case the two optima will be different - we shall observe a "duality gap". In the example, the minimum will be 30 at $v_1=0$, $v_2=3$, while the maximum remains 24.

References

1. Dantzig, G.B. and Fulkerson, D.R. (1956) On the Max-Flow Min-Cut Theorem of Networks. In **Linear Inequalities and related Systems**. Annals of Math. Study No.38. Princeton UP.
2. Ford, L.R. and Fulkerson, D.R. (1962) *Flows in networks*. Princeton UP.
3. Moss Th. (1755) In The Ladies'Diary, or Woman's Almanack. 47.
4. Wesolowsky, G.O. (1993). *Location science*, Vol.1, 5-23.
5. Kuhn, H.W. (1976) Nonlinear Programming. SIAM-AMS Proceedings. Vol.IX. Amer. Math. Soc.

5

Search, Pursuit and Rational Outguessing

5.1 is mainly about the location of a target manoeuvring randomly in order to confuse a possible pursuer. To define the statistics of such a target's position is a crucial step in the definition of effective search tactics. 5.2 concerns a search for buried treasure; in this case the searcher is supposed to have buried the treasure himself, but not to remember exactly where: what does he do under conditions of time and resource constraints to maximize the probability of location? 5.3 is a search problem with an interesting queueing counterpart discussed in 6.3. 5.4 and 5.5 discuss problems of intercepting a target when travelling on a straight line or a circular arc. 5.7 shows how almost the simplest of detection problems, using the most elementary detection law, can entail an incursion into the theory of Bessel functions. 5.6 is different from the others in that it is not overtly connected with search, but rather with what a commander on a battlefield might deduce about his enemy from knowledge of his prior performance.

5.1 HIDE AND SEEK

If you are exposed to detection over a long period, one way of trying to evade a potential pursuer is to adopt evasive tactics. Here we shall look at types of random motion as a model for the realisation of this aim. The searcher's purpose is to make statements in probability terms about where the target may be at a given moment. This enables a search strategy to be formulated. To begin with we shall consider random movement by the target on a straight line segment. This is appropriate to a situation where the searcher is identified with a smuggler attempting to cross a barrier patrolled by customs officers, now identified as the target. The knowledge provided by the probability model may procure a passage with little risk. And from the customs officer's point of view a random patrol may make it difficult for the smuggler to choose a safe point for penetration.

5.1.1 Random Motion on a Straight-Line Segment

For the straight-line segment the x-axis AB from, say, $x=-D$ to $x=+D$ is chosen. The target starts at the origin. First he chooses the direction of the first leg: to the right with probability p; to the left with probability $q(=1-p)$. Next he chooses the length Y of the leg, in the case we shall study first according to the widely applicable exponential distribution with mean a. Thus

$$P[Y<x]=1-e^{-x/a}. \tag{5.1.1}$$

He sets off in the direction prescribed and moves until the distance Y has been covered. If, during the leg, he meets the boundary, i.e. one of the end-points of the line segment, his path is reflected and continues in this way until the whole distance Y has been traversed. He then flips his randomiser to choose the direction and length of the next leg and continues. The problem we shall solve first is the form of the probability $u_n(x)dx$ that at the end of the n-th leg the x-coordinate X_n lies in the small interval $(x,x+dx)$. The fundamental building block is the probability $g(x_1,x)dx$ that, given that the target starts at the point with coordinate x_1, he finishes the next leg in the interval $(x,x+dx)$. It is convenient to distinguish between the two cases:

(a) $-D<x_1<x$, when we shall use a distinguishing superscript, thus

$g^{(-)}(x_1,x)$;

and

(b) $x<x_1<D$, when we shall use $g^{(+)}(x_1,x)$. The reader will find it a nice exercise, also sketched in the Notes, to show that

$$g^{(-)}(x_1,x)=\frac{\cosh(\frac{D-x}{a})}{a\sinh(\frac{2D}{a})}\{pe^{\frac{x_1+D}{a}}+qe^{-\frac{x_1+D}{a}}\}. \tag{5.1.2}$$

By symmetry, $g^{(+)}(x_1,x)$ is obtained from the above by interchanging p and q,

and replacing x and x_1 by their negatives. Thus,

$$g^{(+)}(x_1,x)=\frac{\cosh(\frac{D+x}{a})}{a\sinh(\frac{2D}{a})}\{pe^{\frac{-(D-x_1)}{a}}+qe^{\frac{D-x_1}{a}}\}. \tag{5.1.3}$$

We can use (5.1.2) and (5.1.3) to formulate an integro-difference equation for $u_n(x)$, namely

$$u_{n+1}(x)=\int_{-D}^{x}u_n(x_1)g^{(-)}(x_1,x)dx_1+\int_{x}^{D}u_n(x_1)g^{(+)}(x_1,x)dx_1. \qquad (5.1.4)$$

The first term covers the case where, at the end of the n-th leg, the target is to the left of the point with coordinate x, and the second term is where the end of the previous leg was to the right of x. (5.1.4) can with some labour be applied to generating $u_n(x)$ for $n\geq 2$ starting with $u_1(x)$. If the target starts at the origin, the latter is $g^{(-)}(0,x)$ for $0\leq x\leq D$ and otherwise $g^{(+)}(0,x)$.

Of particular interest is $u^*(x)=\lim u_n(x)$ as $n->\infty$, assuming that it exists (a not unreasonable supposition). It is shown in the Notes that

$$u^*(x)=\frac{(p-q)e^{(p-q)\frac{x}{a}}}{2a\sinh\{\frac{(p-q)D}{a}\}}. \qquad (5.1.5)$$

When $p=q$, so that the target chooses to move right and left with equal probability,

$$u^*(x)=\frac{1}{2D}, \qquad (5.1.6)$$

which means that the target is equally likely to be anywhere on AB so that the searcher has no particular preference for a location to search in first. If D is fixed, while $a->\infty$, (5.1.5) again reduces to (5.1.6), whatever p.

Table 5.1.1 Long-run probability that end of leg lies in left-hand segment

D/a	p=0.1	p=0.5	p=0.9
0.1	0.5199	0.5000	0.4801
0.5	0.5987	0.5000	0.4013
1	0.6900	0.5000	0.3100
2	0.8320	0.5000	0.1680
5	0.9820	0.5000	0.0180
10	0.9997	0.5000	0.0003

Table 5.1.1 puts flesh on these bones. It gives the long-run probability that the target at the end of a leg is in the left half of the segment, i.e. $-D \leq x \leq 0$, for various p and D/a. If $p=0.1$ the searcher should spend most of his time in the left hand segment, but, a smuggler, wishing to avoid detection by the linear patrol, would be best advised to choose to cross right of centre.

Next we look at an alternative distribution for step-length such that explicit and instructive formulae can be obtained. This is the normal on $(0,\infty)$ with density $f(x)$ given by

$$f(x)=\frac{2}{\sigma\sqrt{2\pi}}e^{\frac{-x^2}{2\sigma^2}}, \qquad (0 \leq x \leq \infty).$$

In this case the line-segment will become infinite, modelling a very long line of target uncertainty. Thus there will be no bouncing off terminal points and the analysis becomes easier. Also, in practice, it makes sense only to take $p=1/2$, though we shall include the general case. The formula for $u_{n+1}(x)$ is

$$u_{n+1}(x)=p\int_{-\infty}^{x} u_n(x_1)f(x-x_1)dx_1+q\int_{x}^{\infty} u_n(x_1)f(x_1-x)dx_1. \qquad (5.1.7)$$

By induction it can be shown, when $p=1/2$, that

$$u_n(x)=\frac{e^{\frac{-x^2}{2n\sigma^2}}}{\sigma\sqrt{2\pi n}}. \qquad (5.1.8)$$

We see that as $n->\infty$, the density becomes more and more widely spaced and tends to an increasingly thinly spread uniform density. The same happens as $\sigma->\infty$. This behaviour is consistent with that observed in the study of the finite line segment.

5.1.2 Random Target Motion in Space

The exponentially driven linear target motion studied above has an interesting generalisation into two dimensions. It is an instance of the "random flight" problem propounded first by Rayleigh [2] in an investigation of acoustic propagation ("applied", rather than "applicable", mathematics?) but also associated with the name of the statistician Karl Pearson [3]. It is also, somewhat picturesquely, called the problem of the drunkard's walk, for reasons which will be apparent without explanation. In our terminology it concerns a target which moves, as in one dimension, in a sequence of straight legs, but now oriented to one another at random angles drawn from a uniform distribution on $(-\pi,\pi)$. The searcher wishes to make probability statements about the target's distance R_n from the starting point O, and its direction Θ_n, at the end of the n-th leg.

Although the earlier analysis of the drunkard's walk allowed for legs of different lengths, none attributed a probability distribution to them and that is where the present analysis, based on [1], makes an advance. The long-run result is, of course, predictable by the Central Limit Theorem.

To find the probability distribution of the position of the target at the end of the n-th leg we follow the exposition in [4],pp.419-420, itself based on a brilliant device due to Kluyver [5] employing a discontinuous integral. It is worth while sketching the outline here, leaving details to the Notes.

The situation is illustrated in Figure 5.1.1.

Figure 5.1.1 Final Steps in "Random Flight Path"

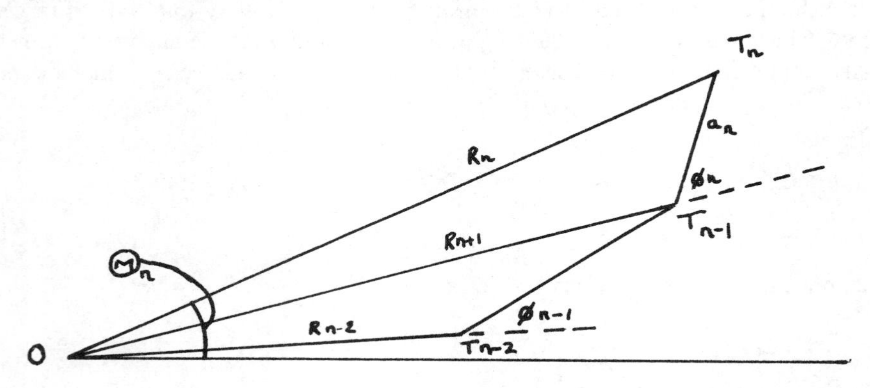

The target begins at O and proceeds in a series of legs as described above. T_n denotes the terminal point of the n-th leg. The direction Φ_n of the n-th leg $T_{n-1}T_n$ with respect to OT_{n-1} is selected from the uniform distribution $U(-\pi, \pi)$. The lengths of the first n successive legs are the components of the vector $\underline{a}_n = (a_1, a_2, \ldots, a_n)$ and R_n is the length of this vector. We seek the probability $P(R_n < r \mid \underline{a}_n)$ that the distance OT_n does not exceed r, given the n leg lengths a_i. Notice that we are not yet requiring that the elements of $\underline{a}_n$ are drawn from a probability distribution. The two key results needed on the way are:

(i) a discontinuous integral ([5], 13.42)

$$r\int_0^\infty J_1(rt)J_0(bt)\,dt = 1 \qquad \textit{for } b<r \tag{5.1.9}$$

$$= 0 \qquad \textit{for } b>r;$$

(ii) and the integral ([5],11.41)

$$\int_{-\pi}^{\pi} J_0(\sqrt{Z^2+z^2+2Zz\cos\phi})d\phi=2\pi J_0(Z)J_0(z). \tag{5.1.10}$$

To lead up to the general result, take $n=3$. Then

$$P(R_3<r|\underline{a_3})=\frac{1}{(2\pi)^3}\int_{-\pi}^{\pi}\int_{-\pi}^{\pi}\int d\phi_3 d\phi_2 d\phi_1$$

taken over all Φ_1 , Φ_2 in $(-\pi,\pi)$ and Φ_3 such that $R_3<r$. The crucial step is to convert the integral with respect to Φ_3 to an integral over $(-\pi,\pi)$ by inserting (5.1.9) in the form

$$r\int_0^{\infty} J_1(rt)J_0(R_3t)dt,$$

so that

$$P(R_3<r|\underline{a_3})=\frac{r}{(2\pi)^3}\int_{-\pi}^{\pi} d\phi_1\int_{-\pi}^{\pi} d\phi_2\int_{-\pi}^{\pi} d\phi_3\int_0^{\infty} J_1(rt)J_0(R_3t)dt.$$

Now

$$R_3^2=R_2^2+a_3^2+2R_2a_3\cos\phi_3$$

and so, changing the order of integration and using (5.1.10), we get

$$P(R_3<r|\underline{a_3})=\frac{r}{(2\pi)^2}\int_{-\pi}^{\pi} d\phi_1\int_{-\pi}^{\pi} d\phi_2\int_0^{\infty} J_1(rt)J_0(a_3t)J_0(R_2t)dt.$$

Again,

$$R_2^2=R_1^2+a_2^2+2R_1a_2\cos\phi_2,$$

so, repeating the previous steps gives

$$P(R_3<r|\underline{a_3})=\frac{r}{2\pi}\int_{-\pi}^{\pi}d\phi_1\int_0^{\infty}J_1(rt)J_0(a_3t)J_0(a_2t)J_0(a_1t)dt$$

since $R_1=a_1$. Finally, integrating out with respect to Φ_1 leaves

$$P(R_3<r|\underline{a_3})=r\int_0^{\infty}J_1(rt)\prod_{m=1}^{3}J_0(a_mt)dt,$$

and the generalization to

$$P(R_n<r|\underline{a_n})=r\int_0^{\infty}J_1(rt)\prod_{m=1}^{n}J_0(a_mt)dt \qquad (5.1.11)$$

is straightforward.

If leg-lengths a_i are independently and identically distributed with probability density function $k(x)$ $(0<x<\infty)$, it follows that at each stage in building up the deconditioned probability $P(R_n<r)$ the product in (5.1.11) must be replaced by

$$\{\int_0^{\infty}k(x)J_0(xt)dx\}^n,$$

giving the unconditional probability

$$P(R_n\leq r)=r\int_0^{\infty}J_1(rt)\{\int_0^{\infty}k(x)J_o(xt)dx\}^n dt. \qquad (5.1.12)$$

When the leg-length distribution is exponential with mean a, $k(x)=\exp(-x/a)/a$, and after some Bessel function manipulations
we arrive at the exact expression

$$P(R_n>r)=2\frac{(\frac{r}{2a})^{\frac{n}{2}}}{\Gamma(\frac{n}{2})}K_{\frac{n}{2}}(\frac{r}{a}), \qquad (5.1.13)$$

which generalises the linear analysis and gives a spatial result. In this formula $K_m(x)$ is Macdonald's modified Bessel function of the third kind (see [4], p.78 for definitions, and *passim* for K_m formulae). The K_m -function is awkward to calculate so we give no tables of the probability for fixed n. But it is important to remark that there are theoretical grounds for supposing that as $n\to\infty$ the distribution tends to the so-called Rayleigh, which is such that $r^2/2$ is exponentially distributed. This is shown in the Notes. Here we remark that as $n\to\infty$,

$$p(R_n>r)\to e^{\frac{-r^2}{2n\sigma^2}}$$

$$\text{where}$$

$$\sigma=a,$$

$$E[R_n]\sim\sigma\sqrt{\frac{\pi n}{2}},$$

$$Var[R_n]\sim n\sigma^2(2-\frac{\pi}{2}).$$

These formulae hold for reasonable arbitrary densities $k(x)$, but σ has to be replaced by the appropriate value. In any case, the area of uncertainty presented to the searcher tends to be an expanding circle for any probability level: for instance, in the case of the exponential leg-length distribution, the 50% zone is a circle with radius $a\sqrt{2n\ln 2}$. "50% zone" here means that the probability is 0.5 that the target lies inside a circle centred on the origin with this radius.

The angle Θ_n made by OT_n with the x-axis is uniformly distributed on $(-\pi,\pi)$, or $(0,2\pi)$, whichever is more convenient. This is because the sum is reduced to the basic range. Take the case of $\Theta_2=\Phi_1+\Phi_2 \bmod 2\pi$. The joint density $P(d\Phi_1 d\Phi_2)$ is $d\Phi_1 d\Phi_2/(2\pi)^2$ and so, integrating out over Φ_1 we get

$$P(d\Theta_2)=\frac{d\Theta_2}{2\pi}\{\int_0^{\Theta_2}\frac{d\phi_1}{2\pi}+\int_{\Theta_2}^{2\pi}\frac{d\phi_1}{2\pi}\}=\frac{d\Theta_2}{2\pi}$$

so that Θ_2 is also uniform over $(0,2\pi)$. It follows similarly that $\Theta_3=\Theta_2+\theta_3$ is uniform on $(0,2\pi)$, and so on. This does not help the searcher to discern an optimal direction for search, but he does at least know something about the range.

We conclude this section with a formula for the probability that the target lies within distance r from its starting point at time t after it started. We preserve the assumption about exponentially distributed leg-lengths but in a different form. We suppose that the target travels with steady speed V and that the target commander's decision at the end of each leg is the *time* to spend on the next leg. If this is t, it means that the distance travelled will be Vt. The probability density function of t is thus $\exp(-t/T)/T$. The formula is derived in [1] and requires operations with Bessel functions. We give it here for its surprisingly simple form:

$$P(R(t)>r)=\exp\{\frac{-t}{T}+\frac{(t^2-\frac{r^2}{V^2})^{1}/2}{T}\}, \qquad (0\le r\le Vt),$$

$$with \quad P(R(t)=Vt)=e^{\frac{-t}{T}}. \qquad (5.1.14)$$

R(t) is the distance of the target from its initially observed point.The formula is exact and the counterpart of (5.1.13). It is possibly more useful to a searcher who may have observed the target once and not seen it since. [1] may be referred to for some analysis of reflected motion in a circular area.

5.1.3 Identification of a likely search area

Aspects of the analysis of a spatially moving target were developed further in [6]. The principal difference is that in [6] a more plausible distribution for turn angle at the end of a leg is substituted for the uniform. This is the **wrapped normal**, natural for angles and possessed of attributes of the normal. If the reader will imagine taking a unit radius circular cylinder, fixing a point on the circumference of the base as the centre of the distribution, and then wrapping the normal density with zero mean and standard deviation σ around the cylinder, adding the density contributions at each point of the circular base, the resultant composite is the wrapped normal density. Denoting the density by $g(\theta)$ we get

$$g(\theta)=\frac{1}{\sigma\sqrt{2\pi}}\sum_{m=-\infty}^{\infty}\exp\{-\frac{1}{2}(\frac{\theta-2m\pi}{\sigma})^2\}, \qquad (-\pi<\theta<\pi), \qquad (5.1.15)$$

an even periodic function with the Fourier cosine expansion

$$g(\theta)=\frac{1}{2\pi}\sum_{n=-\infty}^{\infty}G(n)e^{-in\theta} \quad and$$
$$G(n)=e^{-\frac{1}{2}n^2\sigma^2}, \qquad (5.1.16)$$

whence

$$g(\theta)=\frac{1}{2\pi}+\frac{1}{\pi}\sum_{n=1}^{\infty}e^{-\frac{1}{2}n^2\sigma^2}\cos n\theta, \qquad (-\pi<\theta<\pi). \qquad (5.1.17)$$

Suppose that the target uses this distribution to define an evasive trajectory, as in the preceding section, each angle of turn being measured with respect to the direction of the preceding leg. Let (X_n,Y_n) be the cartesian coordinates of the end point T_n of the n-th leg. To make precise statements in probability terms about the position of the target at the end of a leg requires knowledge of the joint distribution of X_n and Y_n and this does not seem so easy. But, although X_n and Y_n are correlated random variables, there is theoretical evidence that as $n->\infty$ their joint distribution tends to uncorrelated normality. This is supported by numerical experiment. If the coordinates (x,y) of a point P have joint density

$$\frac{1}{2\pi\sigma_1\sigma_2}\exp[-\frac{1}{2}\{(\frac{x-\mu}{\sigma_1})^2+\frac{y^2}{\sigma_2^{\,2}}\}]$$

then the ellipse

$$(\frac{x-\mu}{\sigma_1})^2+(\frac{y}{\sigma_2})^2=\lambda^2,$$

which has probability p of containing (x,y), is specified by choosing λ such that

$$e^{-\lambda^2/2}=1-p.$$

A computer experiment entailing generating two sets of 500 random walks up to the 10-th leg was carried out and it emerged that for $p=0.5$ the numbers of pairs falling within the prescribed ellipse in one case was 271 and 243, with expected value 250. Many more experiments have supported the approach. Details can be found in [6]. As $n->\infty$ the ellipse tends to a circle. This is a very convenient practical conclusion for the searcher.

Finally it is worth remarking that, if the target is making an evasive passage in a general fixed direction with angles of turn made with respect to the fixed direction, the Central Limit applies and the argument about ellipses becomes exact.

5.1.4 Notes

5.1.4.1. Equation (5.1.2) is the probability density function of the leg length required, taking into account reflections, to move the target from a point on the line AB with coordinate x_1 to another point with coordinate x **lying to the right of** x_1. If the choice of direction is "to the right" (with probability p), the target may move to x in a single step of length x-x_1, or with a single reflection at A in a step of length D-x-x_1+D-x $=2D$-x-x_1, or with a reflection at A and then B with a step of length D-$x_1+2D+D+x=4D+x$-x_1, and so on. If, however, the choice of initial direction is "to the left" (with probability q), there has to be at least one reflection from B. Taking into account the various reflections, the target needs in this case legs of lengths D-$x_1+D+x=2D+x$-x_1, D-x_1+2D+D-$x=4D$-x-x_1, D-$x_1+2D+2D+D+x=6D+x$-x_1, and so on. Thus

$$g^{(-)}(x_1,x)=\frac{p}{a}\{e^{-\frac{(x-x_1)}{a}}+e^{-\frac{(2D-x-x_1)}{a}}+e^{-\frac{(4D-x_1+x)}{a}}+e^{-\frac{(6D-x-x_1)}{a}}+...\}$$
$$+\frac{q}{a}\{e^{-\frac{(2D+x+x_1}{a}}+e^{-\frac{(4D+x_1-x)}{a}}+e^{-\frac{(6D+x+x_1)}{a}}+...\}.$$

Introduction of the sum

$$S=1+e^{\frac{-4D}{a}}+e^{\frac{-8D}{a}}+e^{\frac{-12D}{a}}+...=\frac{e^{\frac{2D}{a}}}{2\sinh(\frac{2D}{a})},$$

enables rapid recovery of (5.1.2). As mentioned in the text, (5.1.3) can be obtained from (5.1.2) by exchanging p and q and replacing x and x_1 by their negatives.

5.1.4.2. If $u_n(x)\text{-}{>}u^*(x)$ as $n\text{-}{>}\infty$, (5.1.4) gives

$$u^*(x)=\int_{-D}^{x}u^*(x_1)g^{(-)}(x_1,x)dx_1+\int_{x}^{D}u^*(x_1)g^{(+)}(x_1,x)dx_1.$$

and if this is differentiated with respect to x the following differential equation results:

$$\frac{du^*}{dx}=u^*(x)\{g^{(-)}(x_1,x)-g^{(+)}(x_1,x)\}$$
$$=\frac{(p-q)u^*(x)}{a},$$

after some elementary manipulation. The solution is

$$u^*(x)=Ke^{\frac{(p-q)x}{a}},$$

and the constant is determined by requiring that the integral over $-D\leq x\leq D$ is unity. This is not a surprising result to a seasoned probabilist, but not all searchers possess that gift.

5.1.4.3. The asymptotic Rayleigh form of (5.1.13) can be obtained directly from the asymptotic form of the component integral

$$\frac{1}{a}\int_0^\infty e^{\frac{-x}{a}} J_0(xu)dx = \frac{1}{(1+a^2u^2)^{\frac{1}{2}}},$$

(see [4], p.384). Thus

$$P(R_n \le r) = r\int_0^\infty \frac{J_1(ru)}{(1+a^2u^2)^{\frac{n}{2}}} du.$$

Put $a=b/\sqrt{n}$. Then, as $n\text{-}>\infty$,

$$(1+\frac{b^2u^2}{n})^{\frac{-n}{2}} \sim e^{\frac{-b^2u^2}{2}},$$

(the $\sim$ sign means that the ratio of the two sides->1 as $n\text{-}>\infty$) and

$$P(R_n \le r) \sim r\int_0^\infty J_1(ru) e^{-\frac{b^2u^2}{2}} du = 1 - e^{-\frac{r^2}{2b^2}},$$

by [4], p.393. Thus, as $n\text{-}>\infty$,

$$P(R_n > r) \sim e^{\frac{-r^2}{2na^2}}.$$

References

1. Conolly,B.W. and Roberts, David. (1987).Random walk models for search with particular reference to a bounded region. European Journal of Operational Research, 28, 308-320.
2. Rayleigh, Lord. (1880). On the resultant of a large number of vibrations of the same pitch and arbitrary amplitude. Phil. Mag.,10,73-78. [See also the later paper by the same author: (1919). On the problem of random vibrations and of random flights in one, two and three dimensions. Phil. Mag., 37, 321-347.]
3. Pearson, K. (1905). The problem of random walk. Nature, 72, 294.
4. Watson, G.N. (1944). A treatise on the theory of Bessel functions. Second edition. Cambridge U.P.
5. Kluyver, J.C. (1906).A local probability problem. Proceedings Section of Science, Koningklijke Akademie van Wetenschappen te Amsterdam, **8**, 341-350.

6. Conolly,B.W. and Roberts, David. (1991).Planar exponential random walk with restricted angles of turn. European Journal of Operational Research, 53, 244-252.

5.2 A TREASURE HUNT: OPTIMAL ALLOCATION OF SEARCH RESOURCES

5.2.1 The usual model for detection when a searcher spends time t looking for a target is as follows. Suppose that the target is known to be in a prescribed area **A**. Let $q(t)$ be the probability that continuous search in **A** has failed to detect the target after search time t, and let $\lambda(t)h+o(h)$ be the probability that a detection occurs in the small time interval $(t,t+h)$, where λ is an index of performance of the detection mechanism as a function of time elapsed since search began.[$o(h)$ is shorthand for a collection of terms which, when divided by h, vanish as $h\text{->}0$.]
This leads to the expression

$$q(t)=\exp\{-\int_0^t \lambda(s)ds\},$$

since it is certain that no detection occurred when no search time has elapsed [$q(0)=1$].

An educated thief puts this theory to the test. Some months ago he stole a quantity of gold and platinum bars and buried them after dark in a cave on a remote seashore. When he decided to retrieve them he discovered that there are many more or less identical caves and that he does not know for sure which of them contains the treasure. However, he manages to narrow the search down to two adjacent caves. But then another complication arises. He learns from his CB radio that the forces of law are pursuing him and that he has an hour only to retrieve the booty. He decides that he can spare 15 minutes for thought and calculation, leaving 45 minutes for the actual search. The question then is: How should the 45 minutes search time best be divided between the two possible locations? He takes paper and fountain pen, and argues as follows.

Let $p_i(i=1,2)$ be the probability that the treasure is in cave i. Then $p_1+p_2=1$. If the **instantaneous detection rate** λ is now a constant independent of time, the conditional probability that search for time t will detect the target/treasure is 1-exp($-\lambda t$). Thus, the overall probability P that the treasure is detected by search for time t in cave 1, and for time T-t in cave 2, is

$$\begin{aligned}P&=p_1(1-e^{-\lambda t})+p_2(1-e^{-\lambda(T-t)})\\&=1-p_1e^{-\lambda t}-p_2e^{-\lambda(T-t)},\end{aligned}$$

where T is the total time available for the search (here 45 minutes). The problem is, given T, p_1, λ, to find $t=t'$ $(0\leq t'\leq T)$ so as to maximise P.

Differentiation with respect to t gives

$$\frac{dP}{dt}=\lambda e^{-\lambda(T+t)}(p_1 e^{\lambda T}-p_2 e^{2\lambda t}).$$

P is a smooth function and its maxima/minima occur for such t' as make the derivative zero. Moreover, the second derivative is negative, from which it follows that the turning point is a maximum. The solution is

$$t'=\frac{1}{2\lambda}\ln(\frac{p_1}{p_2})+\frac{T}{2}.$$

The thief considers that search for an hour in a cave that certainly contained the treasure would yield him a 95% chance of finding it. This enables him to estimate λ from

$$1-e^{-60\lambda}=0.95,$$

giving $\lambda=0.05$, nearly enough. He must now estimate p_1. He notes first with satisfaction that if $p_1=0.5$ his calculation tells him to divide search time equally between the caves. However, he rather favours cave 2 and assigns $p_2=0.95$. This gives $t'=-29.4+45/2=-6.9$. Since this value lies outside the range of permissible values he realises that he is being told to spend the whole 45 minutes searching in cave 2, giving almost 90% chance of success.

As he gets to his feet he is beset by a further doubt, for he glimpses a third cave apparently identical with caves 1 and 2. Clearly this must be included in the search plan. 15 minutes have already passed but he decides that it will be worth while to approach the task scientifically, even if it means a reduction in overall search time. He takes a fresh sheet of paper. Happily his pen still contains ink. With obvious notation, the objective probability function is now

$$P=1-p_1 e^{-\lambda t_1}-p_2 e^{-\lambda t_2}-p_3 e^{-\lambda(T-t_1-t_2)}.$$

Repetition of the standard routine to find values t_1',t_2' to yield a maximum value of P gives

$$t_1'=\frac{1}{3\lambda}\ln(\frac{p_1^2}{p_2 p_3})+\frac{T}{3},\qquad t_2'=\frac{1}{3\lambda}\ln(\frac{p_2^2}{p_3 p_1})+\frac{T}{3},$$

$$t_3'=\frac{1}{3\lambda}\ln(\frac{p_3^2}{p_1 p_2})+\frac{T}{3}.$$

λ is still 0.05, but the thief reassesses p_1=0.04,p_2=p_3=0.48. The calculations have left only 30 minutes for the search. He finds t_1'=-23.13, t_2'=t_3'=26.57. But a negative value is inadmissible, so, since p_2=p_3, he assigns equal time, 15 minutes, for search in caves 2 and 3. He calculates that the overall probability of success is reduced to 0.51. Somewhat disheartened he starts to dig. However, the police arrive early and apprehend him as he unearths the treasure in cave 3. He decides to renounce Operational Research.

5.2.2 An interesting, and practically important, extension to the above sad story concerns the deployment of more than one search resource. Hitherto the resource was a limited time T and the parameter λ described the efficiency of its use. But suppose that λ were influenced by another resource, illumination, for example. On a dark night, the thief would almost certainly search more efficiently with a bright light than with a dim one, so suppose that a total illumination potential X is available, and that with potential x the efficiency of search per unit time becomes λx instead of just λ. In the new application λ has, of course, to be appropriately scaled. Taking the simplest example of two caves, and with the allocation of t and x to cave 1, and T-t and X-x to cave 2, the probability of finding the treasure becomes

$$P=p_1(1-e^{-\lambda xt})+p_2(1-e^{-\lambda(X-x)(T-t)}).$$

With some scaling λ can be put equal to unity. Then we can write $x=X\xi$, $y=Y\eta$, $(0\leq\xi,\eta\leq 1)$, and the new problem becomes

$$\begin{gathered}\textit{Maximise}\ \ P=1-p_1e^{-XY\xi\eta}-p_2e^{-XY\bar{\xi}\bar{\eta}}\\ \textit{with}\ p_2=1=p_1,\ \ \bar{\xi}=1-\xi,\bar{\eta}=1-\eta\\ \textit{and}\ 0\leq\xi,\eta\leq 1.\end{gathered}$$

Our friend, now in prison, decides that applicable mathematics in the context of optimization still holds allure and obtains the Governor's permission to lecture to fellow inmates on his new problem. He tells them that they must find ξ and η so that $\partial P/\partial\xi$ and $\partial P/\partial\eta$ are zero, test whether they satisfy the constraints and, if not, use their common sense to select values that do. He mentions that it has been taken for granted that a maximum exists, and that this should be tested, for example by using second derivatives, or by tabulating P. However, even though reduced by symmetry, this problem is more complicated than the previous one with a single resource, and becomes worse with more resources and more locations to search. In the simplest case zeroising the derivatives gives

$$p_1\eta e^{-XY\xi\eta}=p_2\bar{\eta}e^{-XY\bar{\xi}\bar{\eta}},\quad p_1\xi e^{-XY\xi\eta}=p_2\bar{\xi}e^{-XY\bar{\xi}\bar{\eta}},$$

and these show immediately that the proportion of resources to be allocated to the two search areas for optimal results bear a fixed ratio to each other. Thus we write

$$r=\frac{\eta}{\bar{\eta}}=\frac{\xi}{\bar{\xi}}, \quad \textit{giving } \eta=\xi=\frac{r}{1+r},$$

and then r has to be found from the transcendental equation

$$r=\frac{p_2}{p_1}\exp\{-XY(\frac{1-r}{1+r})\}.$$

We notice that when $p_1=p_2$, $r=1$ is a solution. This is as it should be since it tells the searcher to allocate resources equally when he has no prior idea which area is occupied by the target.

Here are some values of r, $\xi=\eta$ and P for $XY=2$. They were obtained by solving the equation $F(r)=0$ where

$$F(r)=\ln r+XY(\frac{1-r}{1+r})-\ln\frac{p_1}{p_2}.$$

Newton's Method is quite convenient, but no doubt other iterations could be tried.

Values of *P* for a 2-Resource Search in 2 Areas

p_1	r	$\xi=\eta$	P
0.1	0.01601594	0.01576348	0.7704
0.2	0.03936869	0.03787751	0.6750
0.3	0.07727654	0.07173324	0.5781
0.4	0.15375191	0.13326254	0.5196
0.45	0.24040909	0.19381436	0.4601
0.5	1	0.5	0.3935

An interesting exercise for the sceptic is to test whether, for given p_1, a "better" value of P can be obtained by varying ξ. Also, why does the optimal P decrease as p_1 increases? There should be symmetry, so P for values of p_1 exceeding 0.5 should be predictable from the above. Is there?

Now generalise to N areas and M resources!

Our friend, by the way, was so captivated by these problems that he used the rest of his sentence to prepare a Ph.D. dissertation on **Search Theory**.

5.3 THE LEAD IN A CHASE

The topic discussed here belongs properly to search theory, but possesses also an interpretation as the model for a type of interactive queueing system which will be discussed in **6.3**.

Figure 5.3.1

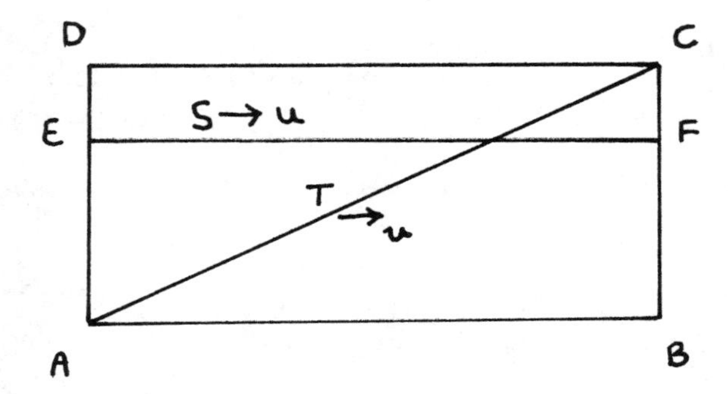

Suppose that a target T is known to be making passage at constant speed along the diagonal AC of a rectangle ABCD. See the Figure 5.3.1. The speed component of the target in the x-direction (i.e. the direction AB) is known to be v. A searcher S patrols the line EF, with the intention of intercepting, if possible, T when it reaches EF. Unfortunately the information available to S is limited. He is given a fix on the x-coordinate of T at random time intervals only. When he is given that information he moves along EF with speed u to try to catch up the projection of T on EF. If he reaches that point before getting the next fix he must stop there. At time $t=0$, S is at E and T is at A. Let t_n $(n=1,2\ldots)$ denote the sequence of instants at which a fix on T is given to S, and let $x_S(t_n)$ and $x_T(t_n)$ denote the x-coordinates of S and the projection of T on EF at time t_n. Then $x_S(0)=x_T(0)=0$. According to the rules:

$$x_S(t_{n+1}) = x_S(t_n)+u(t_{n+1}-t_n) \text{ if } x_S(t_n)+u(t_{n+1}-t_n)<x_T(t_n)(=vt_n);$$

$$x_S(t_{n+1})=x_T(t_n) \text{ if } x_S(t_n)+u(t_{n+1}-t_n)>x_T(t_n)(=vt_n).$$

Denote the *lead* at the time of the n-th fix by $y_n=x_T(t_n)-x_S(t_n)$, and let $\tau_{n+1}=t_{n+1}-t_n$.

Then if

$$(i)\ y_n \geq u\tau_{n+1},$$

$$\begin{aligned} y_{n+1} &= x_T(t_{n+1}) - x_S(t_{n+1}) \\ &= x_T(t_n) + v\tau_{n+1} - x_S(t_{n+1}) \\ &= x_T(t_n) + v\tau_{n+1} - x_S(t_n) - u\tau_{n+1} \\ &= y_n - u\tau_{n+1} + v\tau_{n+1}; \end{aligned}$$

and if

$$(ii)\ y_n \leq u\tau_{n+1},$$

$$\begin{aligned} y_{n+1} &= x_T(t_{n+1}) - x_S(t_{n+1}) \\ &= x_T(t_n) + v\tau_{n+1} - x_T(t_n) \\ &= v\tau_{n+1}. \end{aligned}$$

These may be summarised as

$$y_{n+1} = \max[y_n - u\tau_{n+1} + v\tau_{n+1}, v\tau_{n+1}].$$

Diagrams are a useful guide.

(i)

$\leftarrow y_n \rightarrow B_n \leftarrow v\tau_{n+1} \rightarrow B_{n+1}$

$A_n \leftarrow u\tau_{n+1} \rightarrow A_{n+1} \leftarrow y_{n+1} \rightarrow$

Let $f_n(y)$, $g(t)$ $(y>0, t>0)$ be the probability density functions of y_n and τ_{n+1}. The formulation shows them to be independent, so their joint probability element is the product of the individual elements:

$$dP(y_n, \tau_{n+1}) = f_n(y_n)dy_n\ g(\tau_{n+1})d\tau_{n+1}.$$

Now substitute

$$y_n = y_{n+1} + u\tau_{n+1} - v\tau_{n+1}$$

to get

$$dP(y_{n+1}, \tau_{n+1}) = f_n\{y_{n+1} + (u-v)\tau_{n+1}\}g(\tau_{n+1})dy_{n+1}d\tau_{n+1}.$$

The contribution to $f_{n+1}(y_{n+1})$ is thus the integral of this over

the permissible range of τ_{n+1}, given y_{n+1}, which we see from the diagram is

$$0 < v\tau_{n+1} < y_{n+1}$$

because B_n has to be to the right of A_{n+1}. This gives

$$\int_0^{y/v} f_n\{y+(u-v)t\}\ g(t)\ dt$$

as the contribution to $f_{n+1}(y)$ from case (i).

(ii)

$\leftarrow y_n \rightarrow B_n \qquad \leftarrow v\tau_{n+1} \rightarrow B_{n+1}$

$A_n \leftarrow u\tau_{n+1} \rightarrow A_{n+1}$

In this case the geometric condition is that B_n lies to the left of A_{n+1}, and $y_{n+1}=v\tau_{n+1}$.

Hence

$$\{\int_0^{u\tau_{n+1}} f_n(y_n)dy_n\}g(\tau_{n+1})d\tau_{n+1} \ \ with\ \ \tau_{n+1}=y_{n+1}/v,$$

giving the contribution

$$\frac{g(y/v)}{v}\int_0^{uy/v} f_n(x)dx$$

to $f(y)$.

In the actual problem, the information was supposed imparted to S at exponentially distributed time intervals with mean $1/\lambda$. Thus $y(t)=\lambda\exp(-\lambda t)$, and we get the following integral equation for the density $f_{n+1}(y)$:

$$f_{n+1}(y)=\frac{\lambda}{v}e^{-\frac{\lambda y}{v}}\{\int_0^y e^{\frac{\lambda z}{v}}\ f_n(z+\frac{y-z}{r})dz+\int_0^{y/r} f_n(z)dz\}$$

$$where\ \ r=\frac{v}{u}.$$

This formidable-looking equation turns out to have an elegant-looking solution:

$$f_n(y) = \frac{\lambda}{v}\sum_{m=1}^{n-1} g_{nm} e^{-\lambda y s_n / v} \qquad (y>0)$$

with

$$s_n = \sum_{i=0}^{n} r^{-i},$$

$$g_{nm} = (-1)^m r^{(m+1)(n-\frac{m}{2}-1)} \frac{s_{n-1}s_{n-2}\ldots s_{n-m-1}}{s_0 s_1 \ldots s_{m-1}}.$$

The reader might, with some justification, comment that the connection between this theoretical problem and search seems remote, to say the least. Nevertheless, the problem did arise in a natural way in a study to do with satellite reconnaissance of an area known to contain hostile transiting vehicles and the best methods for intercepting them. The search and evasion tactics were in reality much more exotic than this prosaic example suggests, so much so as to defy attempts in general to calculate probabilities of closest distance of approach, an obvious measure with which to judge the success of a search tactic. Failing this, the analysts turned to computer simulation, a worthy tool but, alas, often fallible. A wrong line here, a quirk in a random generator there, and a false conclusion follows. The more complex a program, the greater the need for proper checks, and what better than an analysable instance that can be compared with program output. That was the first purpose of this work. An added bonus is its interpretation, alluded to at the beginning, as an example of an *interactive* queueing model and we shall look at this in the queueing essay **6.3**.

Reference

Conolly, B.W. (1967) Checking the simulation of a queueing-type situation. Proc. NATO Symposium on Queueing Theory, Lisbon,1965.English Universities Press, 161-166.

5.4 TRACKING AND PURSUIT

5.4.1 We are moving on a straight line track (or *course*) with constant speed, and we notice a *target* in a certain direction. This direction, an angle measured with respect to a fixed line, is technically known as a *bearing*. We know that the target is moving at constant but unknown speed on a straight line whose direction is also unknown. We have no equipment to measure distances from our own position; only bearings. We wish to locate the target using angular measurements only. We shall show, by using a basic property of parabolas, that this is possible.

Because both we (the "searcher") and the target move on straight tracks with constant (and,in general, different) speeds, the ratio of the distances travelled are the same during any time interval. Now it is a characteristic feature of a parabola that if we take any three distinct tangents and intersect them by a fourth tangent, say at points A,B,C, then the ratio of the segments AB/BC is the same, whatever the fourth tangent. (See Notes below).

It follows that all bearing lines ***as well as our track*** are tangents to the same parabola, **and so is the target's track.**

To define a parabola requires four independent pieces of information. For example, in a cartesian frame, the general equation can be written

$$(y\text{-}gx+gX\text{-}Y)^2 = 4\ell(y+x/g\text{-}Y\text{-}X/g,$$

where (X, Y) are the coordinates of the vertex, g is the slope of the axis of symmetry, and 4ℓ is the latus rectum. The four pieces of information can be the definition of four tangents. Since we know that our (i.e. the searcher's) track is a tangent, we need just three bearing lines to provide complete knowledge of the parabola. No matter how many more bearings are measured, provided that the kinematic geometry stays the same, there is no new information to be gleaned about the parabola. In particular, the searcher is unable to identify which tangent corresponds to the target's track.

This problem can be resolved if the searcher changes course and repeats the procedure, taking a minimum of two more bearings on the target, which we assume has not changed course or speed. A new parabola is constructed, and the target's track is a common tangent to the two parabolas.

The procedure is illustrated by Figure 5.4.1. The searcher takes bearings B2, B3, B4 from positions S2, S3, S4, equidistant on the y-axis. The target's corresponding positions are T2, T3, T4. At S4, the searcher changes course and takes further bearings B5 and B6 from positions S5 and S6. The target is then at T5 and T6. The two parabolas P1 and P2 are shown. P1 is defined by the searcher's first track S2S3S4 (up the y-axis), and B2, B3, B4. P2 is defined by B4, B5 and B6, and S4S5S6. B4 and the target's track are common tangents to P1 and P2.

Is there a third, and, if so, what does it mean? In any event, the target's track and whereabouts are identifiable.

Note that Figure 5.4.1 also shows two unnecessary bearing lines which add no information.

In the exceptional case where all bearings are parallel the searcher must either be travelling on a collision course which will intersect that of the target (or has intersected it), or parallel to the target at the same speed.

Figure 5.4.1 The two-parabola diagram

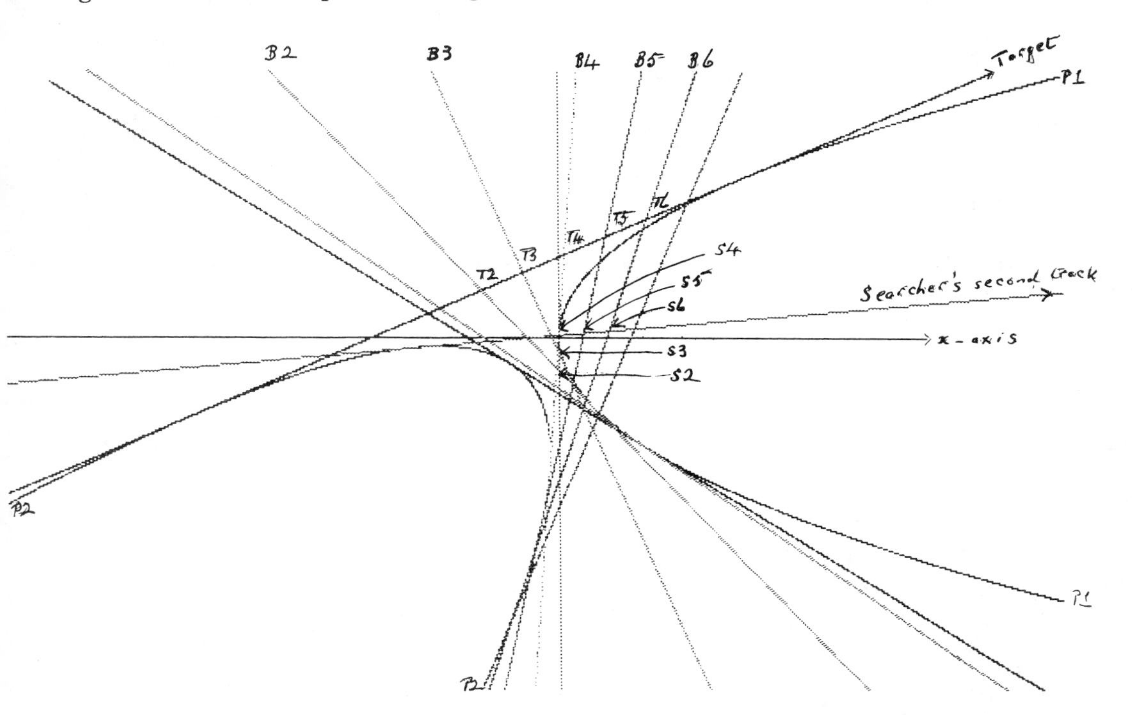

5.4.2. Now that we know the target's position at any time, and consequently also its velocity V_T, say, we may consider intercepting it. Assume the pursuer's, or searcher's (i.e. our own - it is a very rich vocabulary!) velocity to be V_P. The simplest strategy consists of pointing the pursuer's velocity vector always at the target. For an analytical description of the pursuer's path it is convenient to use polar coordinates with the target as origin, by applying the target's velocity vector in the reverse sense. This amounts to discussing the geometry in a space in which the target is placed at rest relative to the pursuer.

In Figure 5.4.2 the target and pursuer are indicated by T and P respectively. r is the distance PT and the polar angle θ is as shown. The kinetic equations describing the motion are

$$\dot{r}=-V_P+V_T\cos\theta, \qquad r\dot{\theta}=-V_T\sin\theta.$$

Figure 5.4.2 Pursuit geometry

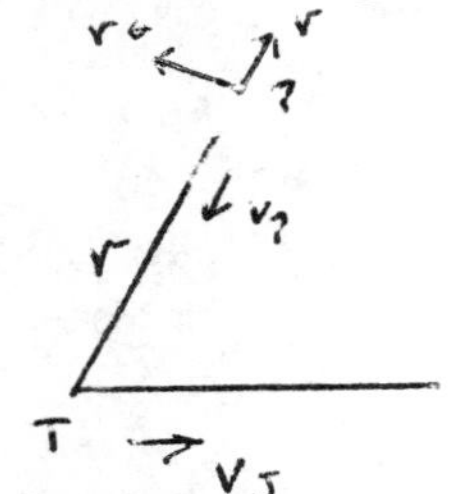

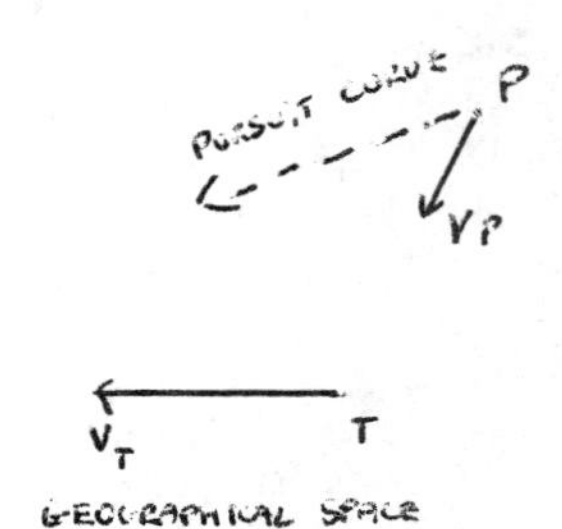

Then, writing $V_P/V_T = \lambda$,

$$\frac{1}{r}\frac{dr}{d\theta}=\frac{d\ln r}{d\theta}=\lambda\operatorname{cosec}\theta-\cot\theta,$$

and

$$\begin{aligned} r&=\exp\{\int(\lambda\operatorname{cosec}\theta-\cot\theta)d\theta\} \\ &=C(\tan\frac{\theta}{2})^{\lambda}/\sin\theta \\ &=C(\sin\frac{\theta}{2})^{\lambda-1}/\{2(\cos\frac{\theta}{2})^{\lambda+1}\} \end{aligned}$$

$$(\textit{since} \int\cot x\,dx=\ln\sin x \textit{ and } \int\operatorname{cosec}x\,dx=\ln\tan\frac{x}{2})$$

where C is a constant of integration.

It is natural to ask whether the pursuer will ever reach the target. Because P's trajectory is a curve (and, naturally, it is known as *pursuit curve*) while that of T is a straight line, it is clear that an actual encounter can happen only if $V_P > V_T$, i.e. $\lambda > 1$.

If this holds, then $r=0$ and $\theta=0$ happen at the same time, and the encounter occurs after time

$$r_0(\lambda+\cos\theta)/\{V_T(\lambda^2-1)\}$$

where r_0 and θ_0 are the initial values of r and θ. For details of the computation see [1].

5.4.3. A slightly more sophisticated approach is that of the *pursuit curve with lead*. In this case the pursuer's velocity vector points ahead of the target. If the lead angle is constant, say ϕ, then

$$\dot{r}=-V_P\cos\phi+V_T\cos\theta, \qquad r\dot{\theta}=-V_T\sin\theta+V_P\sin\phi.$$

Solving these equations for ln r we obtain

$$r=\frac{C\sin\frac{1}{2}(\theta-\psi)^{\mu-1}}{\cos\frac{1}{2}(\theta+\psi)^{\mu+1}}$$

where $\sin\psi=\lambda\sin\phi, \qquad \mu=\lambda\cos\phi/\cos\psi.$

5.4.4 The special case $\sin\theta/\sin\phi=\lambda$ is of some interest. Then $d\theta/dt=0$, and the pursuer moves on a straight-line collision course to the point H where a hit occurs.
Figure 5.4.3 Straight line collision course

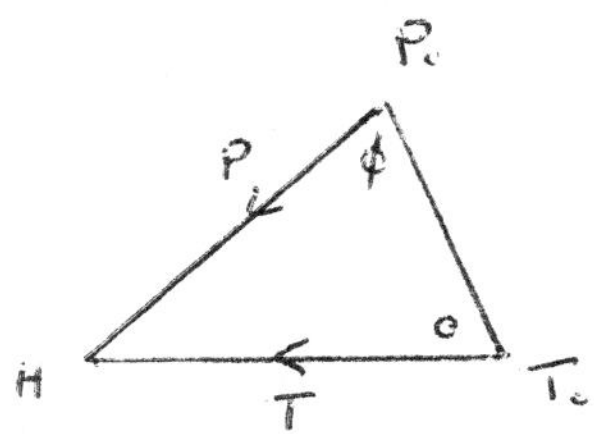

This follows simply from the sine-rule.

If the target's course changes, then the hit-point changes too. The locus of points where the distances from P and from T are in the constant proportion λ is a **Circle of Apollonius** (see Notes. 5.4.6.2).

Another variation is the answer to the question: How may the pursuer try to intercept the target if he knows the initial position of the target and the target's speed V_T, but not its direction? A possibility is that he should move directly towards the target's observed position T_0: if he hits the target (at T_1, say) all is well; if not, he should move on a curve starting at T_1 such that his time along the arc s to the hit-point H is equal to the time the target would take to reach H along the radius vector T_0H. This curve is the *Vignot Spiral*. See Notes 5.4.6.3 for details.

5.4.5 So far we have assumed that the aim of the pursuer is to collide with the target, but it might be the case that he wishes merely to shadow the target at constant distance D. In that case he moves on a ***tractrix***. The pursuer starts at P_0 On the y-axis. The target starts at T_0, the origin, and moves along the positive x-axis, as shown.

Figure 5.4.4 The tractrix

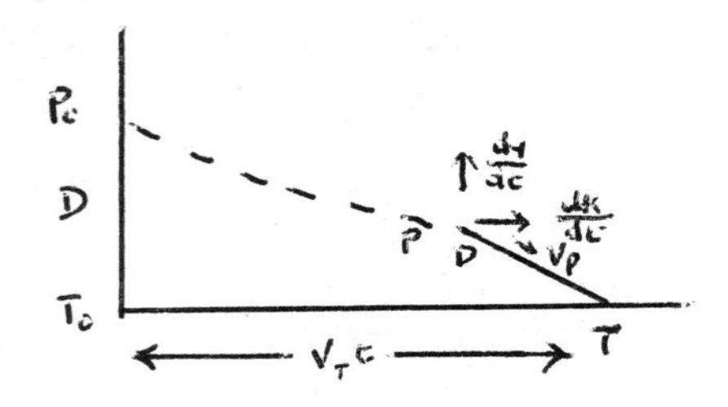

Then

$$V_T t = x + D\cos\psi \qquad (5.4.1)$$

and hence

$$dx/dt = V_T + D(d\psi/dt)\sin\psi. \qquad (5.4.2)$$

Also

$$dx/dt = V_P\cos\psi \qquad \text{and} \qquad dy/dt = -V_P\sin\psi. \qquad (5.4.3)$$

We want to express x and y as functions of the parameter, time t, say. For this purpose we need also

$$V_P = V_T\cos\psi, \qquad (5.4.4)$$

which states that the speed of P relative to T, measured along PT, is zero. Why? Because PT is to remain constant.

From (5.4.2) and (5.4.3) we deduce

$$V_P\cos\psi = V_T + D(d\psi/dt)\sin\psi$$

and then, from (5.4.4),

$$D(d\psi/dt)\sin\psi = -V_T + V_T\cos^2\psi = -V_T\sin^2\psi.$$

Now sin ψ is not zero (see Figure 5.4.4) and it follows that

$$d\psi/dt = -(V_T/D)\sin\psi$$

and hence

$$-V_T t/D = \int_{\pi/2}^{\psi} \frac{du}{\sin u} = \ln\{\tan(\psi/2)\},$$

so that

$$\tan(\frac{\psi}{2}) = e^{(-V_T t/D)}.$$

Since

$$\cos\psi = \frac{1-\tan^2\frac{\psi}{2}}{1+\tan^2\frac{\psi}{2}} = \tanh(\frac{V_T t}{D})$$

we have

$$\sin\psi = sech(\frac{V_T t}{D}).$$

Using (5.4.1), we obtain

$$x = V_T t - D\tanh(\frac{V_T t}{D}), \qquad y = D sech(\frac{V_T t}{D}).$$

5.4.6 Notes

5.4.6.1. Let the parabola, in cartesian coordinates, have equation $y^2=4ax$.
A point P (x,y) on the parabola can be represented in terms of a parameter t by $x=at^2$, $y=2at$. As t varies so P is a variable point on the parabola. The tangent to the parabola at P has equation

$$ty - x - at^2 = 0.$$

Another tangent at P_i $(i=1,2,3)$ has equation $t_i\, y - x - at_i=0$, and the two tangents intersect at the point A_i $(att_i,\ a(t+t_i))$.

The distances A_1A_2 and A_2A_3 satisfy

$$A_1A_2^2 = a^2(t_2-t_1)^2(1+t^2)$$

and

$$A_2A_3^2 = a^2(t_3-t_2)^2(1+t^2)$$

so that their ratios are independent of t, that is of the point P.

The following projective proof may also be of interest both to those with a nostalgic memory for projective geometry and to others who might like to learn more. Let l and l' be fixed straight lines and suppose that a point X moves along l and that another point X' moves along l'. Let their positions be such that if four points C,D,E,F on l **correspond** respectively to the four points C',D',E',F' on l' then the **cross ratios**

$$\frac{EC}{FC}/\frac{ED}{FD} \quad and \quad \frac{E'C'}{F'C'}/\frac{E'D'}{F'D'}$$

are equal. Such **two point ranges** are called **projective.** Now connect any two corresponding points, such as C and C', by straight lines. All such straight lines considered as tangents to a curve generate a conic. Conversely, tangents to a conic intersect any two tangents of the same conic in **projective point ranges.** It follows that, given any five tangents to a conic, the conic is uniquely defined. In particular, the conic will be a parabola if the **points at infinity** of the two ranges on l and l' correspond in the projectivity. Then the line at infinity of the plane is a tangent and the conic will be a parabola. If F and F' are the two points at infinity, then the equality of the cross ratios is

$$\frac{EC}{ED} = \frac{E'C'}{E'D'}$$

which proves the result about the intercepts on the tangents. See [2] for a development of projective geometry.

The encouragement and support of the Defence Research Agency to develop the algebra and programming for Figure 5.4.1 is gratefully acknowledged.

5.4.6.2 Apollonius of Perga (260-200 B.C.) was a teacher of geometry, mainly in Alexandria, but for a short time at the newly established University of Pergamum. The derivation of the conic sections and many of their properties from the same oblique cone with a circular base is due to him.

If the target is at (0,0) and the pursuer at (1,0) then the locus of points with constant ratio of distances from P and T has equation

$$(V_P^2-V_T^2)(x^2+y^2) + V_T^2(2x - 1) = 0,$$

defining a circle. The centre is at

$$\left\{\frac{-V_T^2}{(V_P^2-V_T^2)},0\right\}$$

and the length of its radius is

$$\frac{V_P V_T}{(V_P^2-V_T^2)}.$$

5.4.6.3. The reader may draw a figure as follows. Let T_0 be the origin and T_0P_0 the x-axis, with P_0 the point (1,0). The point T_1 at which interception would occur if T and P moved along the x-axis towards each other has coordinates $(\lambda/(1+\lambda),0)$, where λ is the speed ratio V_T/V_P. If interception does not occur, then P must move along a curve, each point X of which has the property $\lambda s=r$, where s is the arc length along P's trajectory, and $r=T_0H$. There is an ambiguity since P may move either in clockwise or anticlockwise sense and here we choose anticlockwise. The polar angle of H is θ, and since

$$ds^2=dr^2+r^2d\theta^2$$

we obtain the differential equation

$$\frac{1}{r}\frac{dr}{d\theta}=\frac{\lambda}{\sqrt{1-\lambda^2}}$$

(taking the positive sign of the square root), giving

$$r=\frac{\lambda}{1+\lambda}e^{\frac{\lambda\theta}{\sqrt{1-\lambda^2}}},\quad (0<\lambda<1)$$

the Vignot Spiral.

It would be convenient to employ two pursuers in contrary senses.

References

1. Conolly, B.W.(1981) *Techniques of operational research*. Ellis Horwood. Chichester.

2. Veblen, Oswald and Young, John Wesley. (1910). *Projective geometry*. Ginn and Company. Boston, New York, Chicago, London.

5.5 CIRCULAR PURSUIT AND CYCLOIDS

5.5.1 The Circular Pursuit Game

At the Games and Useless Sciences Society (GAUSS) you can play a computer game illustrated by Figure 5.5.1. A dot T on the screen, the **target**, moves along a straight line. The player manipulates another dot P, **the pursuer**, with the purpose of intercepting T. He starts at P_0 and moves along the arc of a circle with radius ρ and, of course, the player's hope is vain if the circle does not intersect T's straight line path.

The velocities v_P and v_T of P and T are known to the player, and there are two versions of the game:

(a) the radius ρ of the circle is fixed and, to win, the player must correctly choose the direction to begin his interception path; this, of course, also means that he has to choose his starting point P_0 in advance, and if no initial direction can be found, the chosen starting point is not feasible;

(b) the initial direction of the path is given and the player must correctly choose the radius of the circular interception path if he can.

Figure 5.5.1 shows a typical geometry which we shall now study in greater detail.

Figure 5.5.1 The Circular Pursuit Game

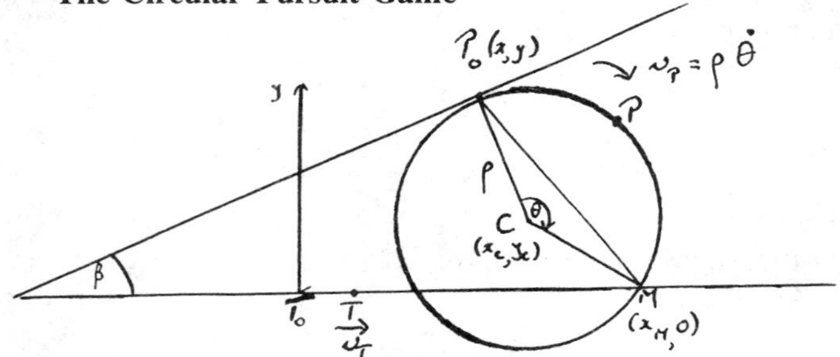

The target T moves to the right along the x-axis and it is convenient to use its starting point as the origin. P_0 is the pursuer's starting point, and β is the initial direction, tangential to the circular path, radius ρ, which, after tracing out an angle θ, meets the target at M. We denote by (x,y) the coordinates of P_0, and by $(x_M,0)$, (x_C,y_C) the coordinates of M and the centre C of the circular pursuit path.

The condition for interception is that the time taken by target and pursuer to reach M are the same, i.e.

$$\frac{x_M}{v_T}=\frac{\rho\theta}{v_P}, \qquad \textit{giving} \qquad x_M=\lambda\rho\theta, \qquad (5.5.1)$$

where λ is the speed ratio v_T/v_P. The distance P_0M is given by

$$P_0M = 2\rho\sin\tfrac{1}{2}\theta. \qquad (5.5.2)$$

Then, the coordinates (x,y) of P_0 are

$$x=x_M-P_0M\cos(\frac{\theta}{2}-\beta), \quad y=P_0M\sin(\frac{\theta}{2}-\beta)$$

or

$$x=\rho\lambda\theta-2\rho\sin\frac{\theta}{2}\cos(\frac{\theta}{2}-\beta)=\lambda\rho\theta-\rho\{\sin(\theta-\beta)+\sin\beta\}$$

$$y=2\rho\sin\frac{\theta}{2}\sin(\frac{\theta}{2}-\beta)=\rho\{\cos\beta-\cos(\theta-\beta)\}. \qquad (5.5.3)$$

These may be written as

$$\frac{x}{\rho}=\lambda(\theta-\beta)-\sin(\theta-\beta)+\lambda\beta-\sin\beta,$$

$$\frac{y}{\rho}=\{1-\cos(\theta-\beta)\}-(1-\cos\beta), \qquad (5.5.4)$$

(5.5.4) are the equations of a *cycloid*, a curve traced out by a point fixed on a rolling circle. This is elaborated in the Notes (**5.5.5**), and a comparison can be made between the equations (5.5.9) derived there directly and (5.5.4) obtained from thr game description. Thus, the locus of starting points P_0 that make P successful is a cycloid.

It is natural to ask what is the relationship between the circle on which the pursuer moves and the circle which generates the cycloid. This turns out to be a circle whose centre C' is the reflection of C in the chord PM. Figure 5.5.2 shows a circle centred at C', with radius ρ, which obviously must pass through P and M.

Figure 5.5.2 The generating circle

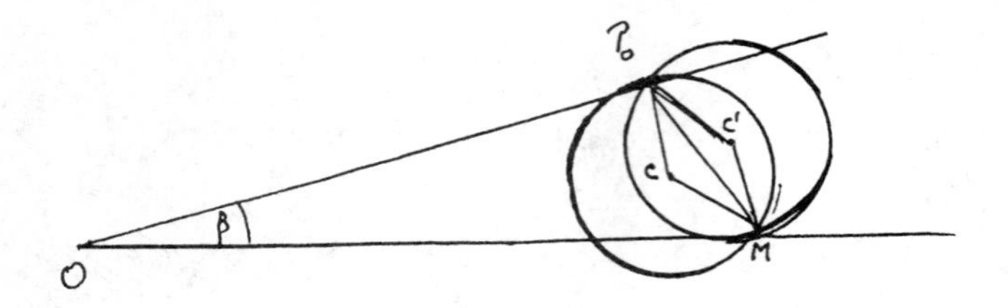

The angle C'MO is equal to $\frac{1}{2}\pi$-β, and does not depend on θ. Hence, wherever the circular trajectory of P lies, the corresponding reflected circle with radius ρ moves in such a way that its centre is always at a distance $\rho\cos\beta$ above the target's track. The supporting algebraic argument is as follows:
The coordinates of C' are

$$x_{C'}=\lambda\rho\theta-\rho\sin\beta, \quad y_{C'}=\rho\cos\beta,$$

and it is easily checked that indeed

$$x_{P_0}=x_{C'}-\rho\sin(\theta-\beta)=\lambda\rho\theta-\rho(\sin\beta+\sin(\theta-\beta))$$

$$y_{P_0}=y_{C'}-\rho\cos(\theta-\beta)=\rho(\cos\beta-\cos(\theta-\beta)),$$

as at (5.5.3).

This confirms that P_0 is generated by a rolling circle with centre at C'. C' moves with speed $dx_{C'}/dt=\lambda\rho d\theta/dt$, and so the radius of the **generating circle** is $\lambda\rho$. The point P which traces out the cycloid is at distance ρ from C', and thus the cycloid is a trochoid if $\lambda>1$, has cusps if $\lambda=1$, and is otherwise prolate. We see also from (5.5.3) that P_0 lies on the circle

$$(x-\rho\lambda\theta)^2+y^2=4\rho^2\sin^2\frac{\theta}{2}, \qquad (5.5.5)$$

with centre $(\lambda\rho\theta,0)$ (that is the point M) and radius $2\rho\sin\frac{1}{2}\theta=MP_0$. Also the centre C of the path of P_0 has coordinates

$$x_C=x_M-\rho\sin(\theta-\beta), \quad y_C=-\rho\cos(\theta-\beta), \qquad (5.5.6)$$

and also lies on a circle, namely

$$(x-x_M)^2+y^2=\rho^2, \tag{5.5.7}$$

with centre at M (x_M,0) and radius ρ. To each value of θ corresponds a different initial point P_0, and different circles (5.5.5) and (5.5.7). We shall look more closely at these geometrical aspects later.

5.5.2 Solution of the Game

The game-player in the arcade can try to solve the two versions of the game by manipulation of the joystick furnished by the GAUSS. But if he objects to paying the entrance fee for each try he may wish to retire to a corner with paper and pencil.

The first version gives the player all the elements in equations (5.5.3) except θ and β. β is asked for, but θ has to be found as well. The second version is similar, but only θ and ρ are unknown. In both cases there are two unknowns to find from two equations: a well-posed problem complicated only by the involved form in which the unknowns appear. The best way to solve them may require some experimentation.

Version (a) can in principle be tackled by finding θ from (5.5.5): then β follows from the more convenient of (5.5.3). The coordinates of the desired starting point P_0 are given. For Version (b) the ratio x/y does not contain ρ, and since β is given, the problem is again just to find θ, whereupon ρ follows from one of (5.5.3). We give an

Example Take $x=3, y=1.75$, $\lambda=1.5$, $\rho=1$. First we try to solve $f(\theta)=0$, with

$$f(\theta)=(x-\lambda\rho\theta)^2+y^2-4\rho^2\sin^2\frac{\theta}{2}$$

$$=(x-\lambda\rho\theta)^2+y^2-2\rho^2(1-\cos\theta).$$

Newton's Method can be tried, viz. the iteration

$$\theta_1=\theta_0-f(\theta_0)/f'(\theta_0)$$

where the dash denotes differentiation. We get

$$f'(\theta)=-2\rho\{\lambda(x-\lambda\rho\theta)+\rho\sin\theta\},$$

and, starting with an initial guess of $\theta_0=0.5$, the following sequence of iterates is obtained: 0.5, 1.5, 1.93, 2.0976, 2.1584, 2.1691, 2.169499 ($=124°.3$). Then, from the y-equation,

$$\sin(\theta/2-\beta)=\frac{y}{2\rho\sin(\theta/2)}$$

we obtain β=-0.3418.

Using the same numerical example for Version (b) of the game, we start with

$$\frac{x}{y}=\frac{\lambda\theta-\sin(\theta-\beta)-\sin\beta}{\cos\beta-\cos(\theta-\beta)}.$$

We now know β, so we try to solve g(θ)=0 with

$$g(\theta)=\lambda\theta-\sin(\theta-\beta)-\sin\beta-\frac{x}{y}\{\cos\beta-\cos(\theta-\beta)\},$$

$$g'(\theta)=\lambda-\cos(\theta-\beta)-\frac{x}{y}\sin(\theta-\beta).$$

Then the iteration

$$\theta_1 = \theta_0 - g(\theta_0)/g'(\theta_0),$$

starting again with θ_0=0.5, homes in on the (genuine) zero θ=0, which the games-player does not want. Some experimentation quickly leads to the other zero, the correct value, and then ρ can be found easily from, say, the *y*-formula (5.5.3).

5.5.3 A Special Case of Circular Pursuit

If in equation (5.5.3) we let $\rho\text{-}>\infty$ and $\theta\text{-}>0$ in such a way that $\theta\rho\text{-}>$a finite value *D*, the pursuer's path P_0M becomes a straight line and the problem degenerates into that of linear pursuit discussed in essay **5.4**. Since $2\rho\sin\frac{1}{2}\rho\text{-}>D$, equations (5.5.3) give for the coordinates of P_0,

$$x = D(\lambda - \cos\beta), \quad y = -D\sin\beta.$$

Put angle $MT_0P_0 = \phi$. Then $x=T_0P_0\cos\phi$, $y=T_0P_0\sin\phi$, that is,

$$D(\lambda-\cos\beta) = T_0P_0\cos\phi, \quad D\sin\beta = -T_0P_0\sin\phi,$$

so that

$$\frac{\cos\beta-\lambda}{\sin\beta}=\frac{\cos\phi}{\sin\phi},$$

or

$$\lambda\sin\phi=\sin(\phi-\beta).$$

The relation between ψ, β and ϕ shown on the adjacent diagram gives

$$\lambda\sin\phi=\sin(\pi-\psi)=\sin\psi,$$

the classical formula for the "aim-off" angle ψ in terms of observed ϕ for linear pursuit.
Diagram

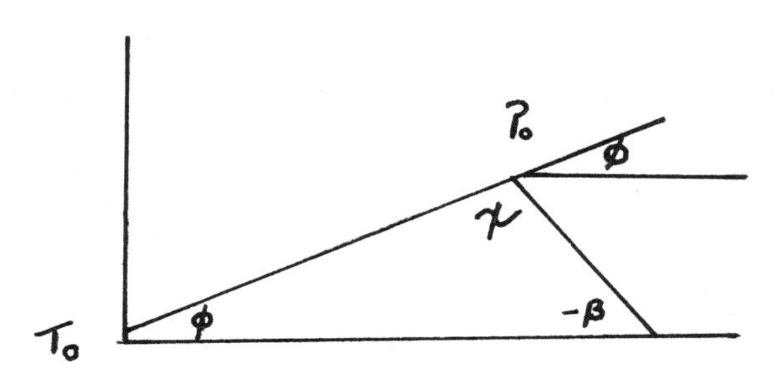

5.5.4 A closer Look at the Geometry

Figures 5.5.3a,b,c show the cycloidal loci of initial points P_0 from which it is possible to hit a moving target by traversing a circular arc. The unit of length is the radius of the circle. The speed ratios are $\lambda=0.5,1,1.5$, respectively, and a glance at the shapes confirm that the curves are prolate (with loops) when $\lambda=0.5$, with cusps when $\lambda=1$, and trochoidal when $\lambda=1.5$.

Each Figure is a plot of the (x,y) coordinates of P_0 for $\beta=0,\pi/2,\pi$. The coordinates were calculated from (5.5.4) with $\rho=1$, namely,

$$x=\theta\lambda-\sin\beta-\sin(\theta-\beta), \quad y=\cos\beta-\cos(\theta-\beta),$$

by varying θ over the range $(0,2\pi)$ for each (λ,β) pair. x increases like $\lambda\theta$ with a periodic component, while y is entirely periodic.

To each coordinate pair (point P_0) correspond values of β and θ, but there are obviously limitations. Moreover, through each point P_0 on the curves passes a circle given by the normalized form of (5.5.5), viz.

$$(x-\lambda\theta)^2+y^2=4\sin^2\theta/2,$$

centred at the point $(\lambda\theta,0)$ of impact of P with the target T. The radius $2\sin\frac{1}{2}\theta$ of these circles is zero at $\theta=0$ and 2π, the corresponding points P being (0,0) and $(2\pi\lambda,0)$. Also the radius has a maximum value of 2. These observations are independent of β.

All the circles, for all possible values of θ, are enclosed by an **envelope**, that is, a curve that touches each of them. It follows that the envelope contains all the points P_0 from which an interception is feasible.

This envelope can be found by the standard procedure of differentiating (5.5.5) in the form $f(\theta)=0$ with respect to θ, equating the derivative to zero, and eliminating θ. We can not quite do that, but the two equations give

$$x=\lambda\theta-\frac{\sin\theta}{\lambda},$$

$$y=\pm 2\sin(\theta/2)(1-\frac{\cos^2\theta/2}{\lambda^2})^{1/2}.$$

The envelope can then be plotted by varying θ and calculating x and y.

Figure 5.5.4 shows the envelope in the case corresponding to Figure 5.5.3b, with $\lambda=1$ and more β-trajectories than before. A comparison with (5.5.4) shows that the upper half of this envelope is the cycloid for $\beta=0$. (N.B. $\rho=1$ and $1-\cos\theta=2\sin^2\frac{1}{2}\theta$.)

It has the two axes of symmetry $y=0$ and $x=\lambda\pi$. If $\lambda\geq 1$, $y=0$ for $\theta=0$ and 2π, corresponding to $x=0$ and $2\pi\lambda$. If $\lambda<1$, (0,0) and $(2\pi\lambda,0)$ are isolated points inside the envelope. $y=0$ holds also when $\cos\frac{1}{2}\theta=\lambda$.

Figure 5.5.3a

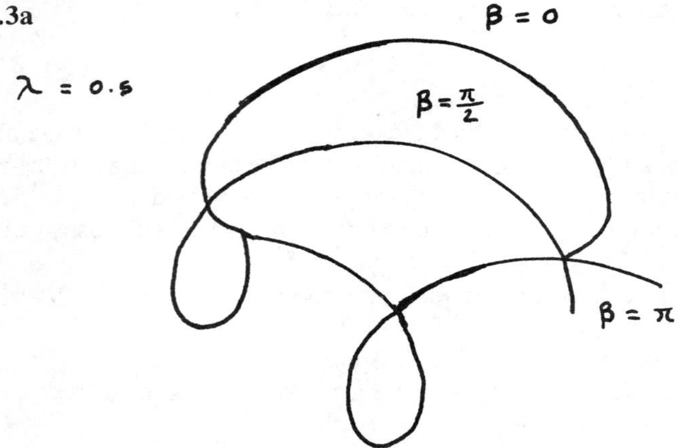

Figure 5.5.3b

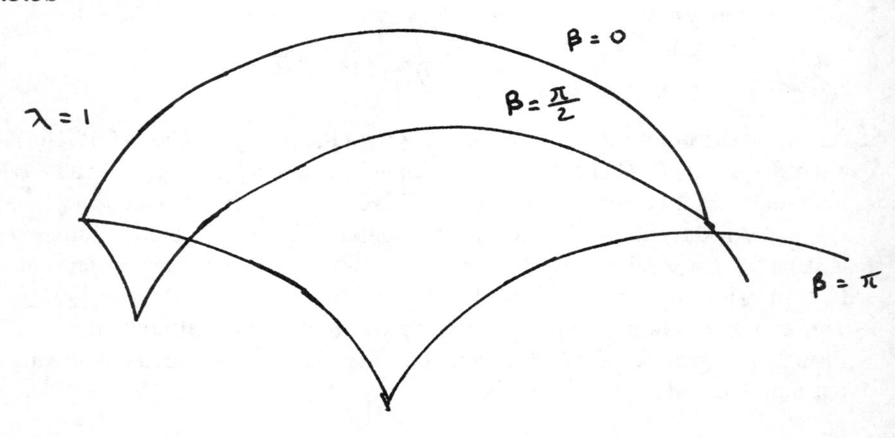

Figure 5.5.3c

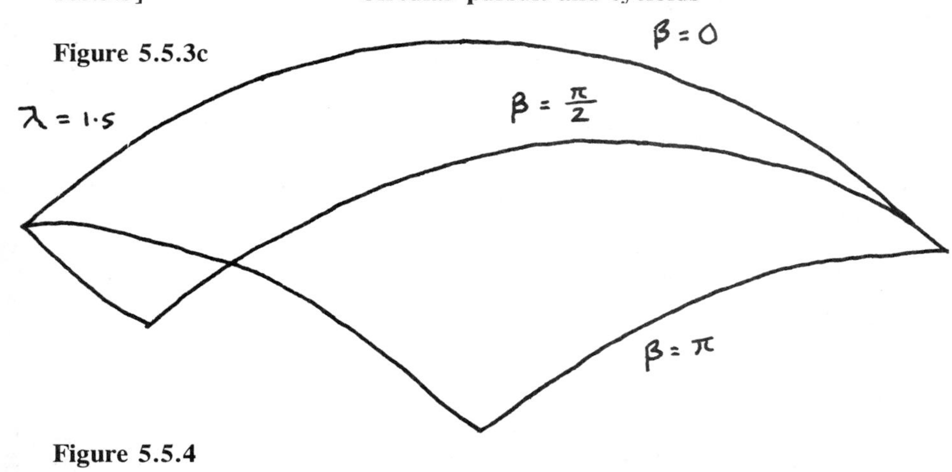

Figure 5.5.4

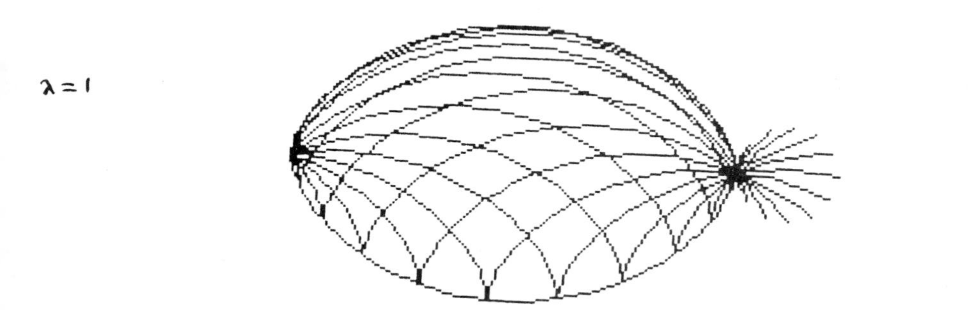

It is not, of course, possible to obtain an interception from an arbitrary point P_0 with arbitrary initial direction β. That is, not just any ray from (0,0) will meet some β-cycloid with any given β.

Let us assume that we are looking for a feasible point from which to start in the direction β_0 on a circle with radius ρ, and to obtain a hit on the target. Is this possible with given λ?

Figures like 5.5.3 a,b,c provide the answer. To fix ideas, let $\lambda=0.5$ (Figure 5.5.3a) which concerns prolate cycloids.

Consider a point (x_0,y_0). If it happens to lie on the β_0 cycloid in Figure 5.5.3a, then we obtain a hit provided that $\rho=1$. But if we insist on $\rho\neq1$, then we need to use a figure different from 5.5.3a obtained by multiplying the coordinates (x,y) with respect to origin (0,0) by ρ. Such a dilation produces the required figure, because when ρ is the unit of length, equations (5.5.4) are non-dimensional.

To find the required point draw the ray through (0,0) and (x_0,y_0) and find the point on this ray whose distance from (0,0) is ρ times that of (x_0,y_0). The new point will be the starting point required, provided it lies on the new (dilated) β_0-cycloid.

A scrutiny of Figure 5.5.3a shows that it might not be possible to find a β_0 for which this is the case. A ray from (0,0) will only meet a β-cycloid a second time, and hence also its dilated copy, if its slope is smaller than that of the tangent at the double point to the cycloid with smallest β which has a double point inside the envelope. The first such double point appears at (0,0), when $x=0$, i.e. $\sin\beta/\beta=\lambda$. From any point with larger slope we can not obtain a hit if we insist on starting angle β_0.

Helpful discussions with Mr. G.B. Trustrum of the University of Sussex are gratefully acknowledged.

5.5.5 Notes

We remind ourselves what cycloids are. They are a family of curves traced out by a point fixed to a circle which rolls without slipping on a straight line. Figure 5.5.5 shows such a circle of radius R with a point P fixed to it at a distance r from the centre C. In this case $r<R$.

Figure 5.5.5 Generation of Cycloids

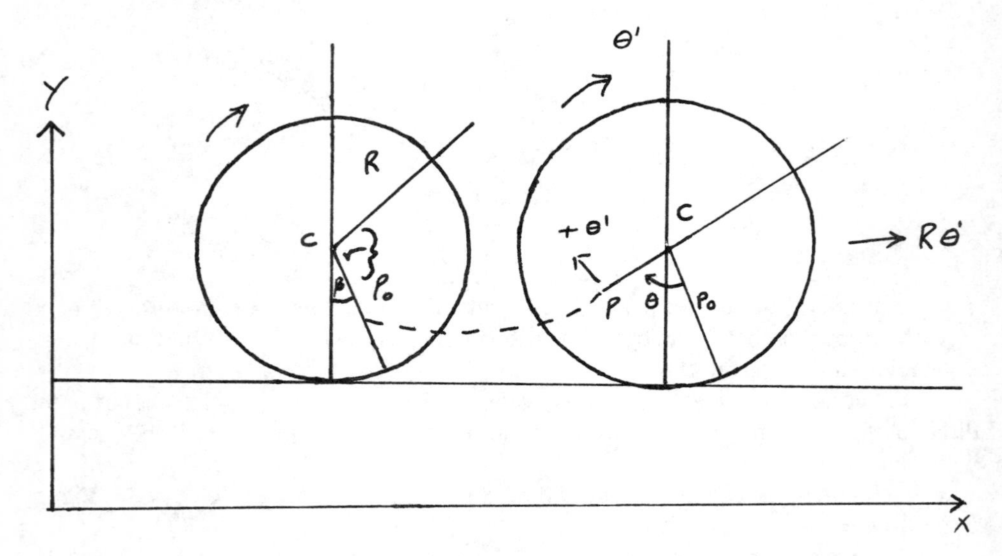

The circle rolls on the line shown towards the right with constant angular velocity θ'. The line may be conveniently thought of as parallel to the X-axis of a cartesian system. The initial position is shown by the circle on the left. P_0 is the initial position of P, and CP_0 makes an angle β with the vertical. The position after time t is shown by the circle on the right. P has moved through an angle θ and its speed component relative to C is $r\theta'$ as shown, at right angles to CP. The prime signifies differentiation with respect to time t. Since the circle is rolling steadily without sliding C has velocity $R\theta'$ parallel to the direction of movement. It is easily seen that the speed components (X', Y') of P's velocity are given by

$$X' = R\theta' - r\theta'\cos(\theta - \beta), \quad Y' = r\theta'\sin(\theta - \beta). \tag{5.5.8}$$

If when $t=0$, $\theta=\beta$, $X=X_0$, $Y=Y_0$, the solution of (5.5.8) is

$$X = R(\theta - \beta) - r\sin(\theta - \beta) + X_0$$

$$Y = r\{1 - \cos(\theta - \beta)\} + Y_0. \tag{5.5.9}$$

(5.5.9) gives the (X, Y) coordinates of a point on a cycloid with the angle θ as parameter. If $r<R$, so that the point P tracing the cycloid is internal to the circle (or wheel), the cycloid is smooth, that is, has no double points, or points having more than one tangent, and is often called a *trochoid*. If $r=R$, the point P is on the rim of the wheel and the cycloid has double points called cusps, with two coincident tangents. If $r>R$, P is external to the wheel, though fixed to it, and the cycloid has double points with two different tangents, giving rise to loops: in this case the cycloid is said to be *prolate*.

The three forms are shown in Figure 5.5.6. The scale can be adjusted so that the equations are in fact

$$x = \theta + r\sin\theta, \qquad y = 1 + r\cos\theta,$$

which imply initial conditions $\theta=0$, $x=0, y=1+r$.

Figure 5.5.6 The Three Cycloidal Forms

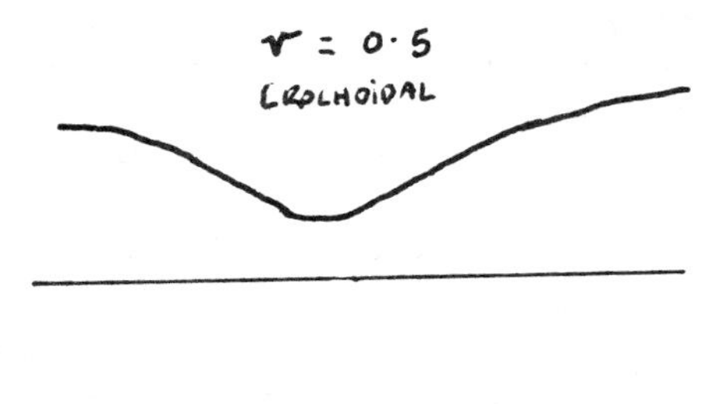

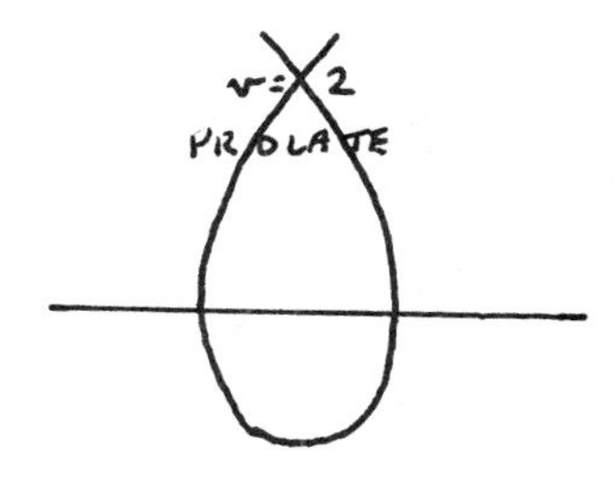

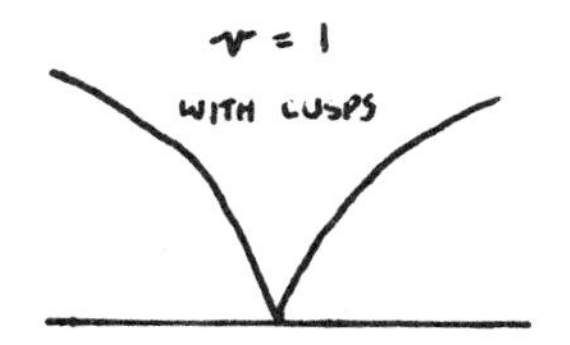

5.6 INVERSE LANCHESTER THEORY

5.6.1 Before embarking it would be better to begin with an explanation. Frederick William Lanchester is probably most widely remembered as an eminent engineer whose career straddles the 19th-20th centuries, as a pioneer in aerodynamics and the design of aircraft and motor car transmissions. But he also started an industry, probably unknowingly, which has flourished since his death. This is best known in that compartment of operational research concerned with military matters: appropriately it is called *Lanchester Theory*. In simplest form this models deterministically the evolution of a battle between two opponents by coupled differential equations whose solutions represent as real variables the force strengths, x, y, say, of the combatants X, Y respectively, as continuous, decreasing functions of time t. The two basic models are:

$$\dot{x}=-\lambda xy, \qquad \dot{y}=-\mu xy, \tag{5.6.1}$$

$$\dot{x}=-\alpha y, \qquad \dot{y}=-\beta x, \tag{5.6.2}$$

and there is a third hybrid

$$\dot{x}=-\alpha y, \quad \dot{y}=-\mu xy, \tag{5.6.3}$$

where dots, as usual, denote differentiation with respect to time. In each case the model holds only as long as x and y remain positive. When one of them becomes zero it is said that the other is the victor. Of course it is possible for them to vanish together, but that is a practically negligible eventuality. The initial conditions are $x=M, y=N$ when time $t=0$. Lanchester [1] described (5.6.1) as a representation of *ancient warfare* which is such that the combatants on either side do not possess the capability of aiming at specific targets in the opposition. Thus, every arrow (or other missile) fired by a Y-unit at a rate of one per unit time is supposed to fall into an area of size A where the density of X-forces is x/A, and has a chance p, say, of destroying just one X-unit. Thus, on average, each attack by a Y-unit slaughters px/A X-units per unit time and, taken over the whole of the Y-force, this gives the rate λxy in (5.6.1). A similar argument gives the companion equation in (5.6.1). (5.6.2), said Lanchester, represents *modern warfare*, where each Y-unit can pick and attack a specific X-unit as target so that the loss rate of X-units is independent of x. The symmetrical argument gives the companion in (5.6.2). The hybrid (5.6.3) is often used to model guerrilla, or terrorist, operations and, as written, assigns to Y the role of guerrillas hidden in their mountain strongholds, able to select particular X-targets, while the X-force is compelled to fire blindly, hoping for the best.

Each of (5.6.1),(5.6.2),(5.6.3) implies an invariant. These are, respectively,

$$\mu x - \lambda y = \mu M - \lambda N = K_l, \tag{5.6.1a}$$

$$\beta x^2 - \alpha y^2 = \beta M^2 - \alpha N^2 = K_s, \tag{5.6.2a}$$

$$\mu x^2 - 2\alpha y = \mu M^2 - 2\alpha N = K_m. \tag{5.6.3a}$$

The linear and quadratic forms of (5.6.1a) and (5.6.2a) are the reason why models (5.6.1) and (5.6.2) are usually referred to as Lanchester's Linear and Square Laws, respectively.

Table 5.6.1 shows the solution of the *modern warfare* model (2) for the initial values $M=75, N=100$ and attrition coefficients $\alpha=0.01, \beta=0.02$. The general who believed Lanchester Theory and the appropriateness of the model for the battle about to be joined, and who, perhaps improbably, knew M, N, α, β, would be able to calculate $K_s=12.5$ and would therefore expect X to be the victor. Table 5.6.1 reinforces this view.

Table 5.6.1 **Evolution of a Square-Law Battle**

Initial values: $M=75; N=100$. Attrition Coefficients: $\alpha=0.01; \beta=0.02$.

t	x	y
0	75	100
10	65.72	85.95
20	57.75	73.63
30	50.94	62.77
40	45.16	53.18
50	40.27	44.65
60	36.20	37.02
70	32.84	30.12
80	30.15	23.84
90	28.06	18.02
100	26.53	12.57
110	25.54	7.37
120	25.05	2.32
125	25.00	-0.18

5.6.2. But what if the general **is** the X-commander and knows only M (his initial strength) and β, the coefficient of attrition that his own weapons are capable of inflicting on the enemy Y ? He calls a staff meeting and explains that he wants to know the outcome of the next battle. What will he do about N and α?

A staff officer who had completed a postgraduate operational research course reminds the general deferentially that he does have more data. There was that battle fought a week ago against the same opposition. Surely he had a record of Y and X casualties. The general agrees but wants to know what that has to do with it. The staff officer asks if he can use the blackboard, and this is what he explains.

(5.6.2a) can be put in the form

$$x^2-\frac{\alpha}{\beta}y^2=M^2-\frac{\alpha}{\beta}N^2,$$

and since Y-casualties z are given by

$$z = N - y,$$

this is equivalent, after a little calculation, to

$$M^2-x^2=\frac{\alpha}{\beta}(2Nz-z^2). \qquad (5.6.4)$$

The observed X-force levels x_1, x_2 at times t_1, t_2, and the Y-casualties z_1, z_2 give two equations from which α/β can be eliminated and N calculated:

$$N=\frac{1}{2}\frac{z_2^2(M^2-x_1^2)-z_1^2(M^2-x_2^2)}{z_2(M^2-x_1^2)-z_1(M^2-x_2^2)}. \qquad (5.6.5)$$

(5.6.4) at $t=t_1$ and t_2 gives α/β, and if X knows β, he can find α. To find out how he stands in the next battle the general has to estimate the next N and then deploy such an M that $\beta M^2-\alpha N^2>0$. This ought to ensure victory!

Example From Table 5.6.1 take observations at $t_1=10$, $t_2=30$. We then have $(x_1,y_1)=(65.718,85.952)$, $(x_2,y_2)=(50.944,62.774)$, giving $(z_1,z_2)=(14.048,37.227)$. Then (5.6.5) gives $N=99.977$ (!) and from (5.6.4),$\alpha/\beta=0.500$, so that $\alpha=0.01$.

If the general is not too sure about his own weapon performance , measured by β, he can use the same data to estimate it. He gets N and α/β as above and then, on the advice of his clever staff-officer, notices that (5.6.2) implies that

$$x \mp y\sqrt{\frac{\alpha}{\beta}} = \{M \mp N\sqrt{\frac{\alpha}{\beta}}\}e^{\pm \omega t}. \qquad (5.6.6)$$

With the above data at t_1 he finds $\omega = \surd(\alpha\beta) = 0.01414$ and deduces $\beta = \surd(\alpha/\beta)\surd(\alpha\beta) = 0.02$. We must confess to the reader that the general is unlikely to have fractional force strengths and if the values in the table are rounded to the nearest integer he might get misleading results. All the more reason to take another viewpoint.

5.6.3 The procedure described above is the solution to a so-called **inverse problem** and for it to be useful the commander must, as we have just observed, have faith in the deterministic model. He might on the other hand take the view that any battle is a random process and that, **on average**, his inverse problem calculations should be "in the right ballpark". We can look into this proposition by computer simulation. That is to say, we pick M, N, α, β and simulate, say, 100 battles. For each battle we calculate estimates of α and β from casualties, as above, and finally look at the distributions of the estimates. Such a procedure using the constants of Table 5.6.1, and based on 100 battles gives the following statistics of the estimates:

$$\text{Mean } (\alpha, \beta) = (0.00983,\ 0.0197),$$

$$\text{Standard deviation}(\alpha, \beta) = (0.0020, 0.0030).$$

So the means appear to home in on the true values, but the variation is possibly alarming. How to run such a simulation is another story and, although it is interesting enough, it would divert us from our main purpose here. So we return to the time independent invariants (5.6.1a), (5.6.2a),(5.6.3a), and ask if they have any probability significance.

5.6.4 It turns out that they do have such significance and this is most clearly seen from the ancient warfare model for reasons that will shortly be apparent. The basis of the argument is to model a battle between X and Y as a sequence of Bernoulli trials with probabilities of success and failure p_{xy}, q_{xy}, respectively. Thus, in general, the success probability depends on the force levels when a trial, or engagement between the opponents, takes place. The elementary theory of Bernoulli trials confines its attention to constant probability of success. But we must first say what we mean by success and failure. We mean here by **success** that in the engagement a Y-unit is lost, and, conversely, by **failure,** that an X-unit is lost. So we are taking the point of view of X. The probability of victory is thus the probability that X destroys all of Y's initial strength of N units before losing all of his own initial strength of M units. Note that a Bernoulli trial has by definition only the two possibilities for its outcome: success or failure, and there is no possibility of a draw.

The elementary theory of Bernoulli trials concentrates on two associated random variables: S_n, the number of successes in n trials, and T_n, the number of trials to the n-th success. S_n has possible values the integers from 0 to n, while T_n may have an infinity of positive integer values from n upwards. We assume that the reader is familiar with the two basic results:

$$P(S_n=m)=\binom{n}{m}p^m q^{n-m},\quad P(T_n=n+m)=p^n\binom{n+m-1}{n-1}q^m. \tag{5.6.7}$$

The second of these can be deduced from the first by noticing that if the outcome of the $(n+m)$-th trial is the n-th success the previous $n+m$-1 trials must have contained n-1 successes, while the last, $(n+m)$-th, has to be a success. The first gives the binomial distribution and the second the negative binomial.

Now suppose that the battle between X and Y is such that the probability of success at each engagement is the constant p. Let $u(M,N)$ be the probability that, with initial values M on the X-side and N on the Y-side, X is the victor. Then, as explained,

$$u(M,N)=P(N\leq T_N\leq M+N-1)=p^N\sum_{m=0}^{M-1}\binom{N+m-1}{N-1}q^m. \tag{5.6.8}$$

A second form follows from the observation that if the N-th success occurs at, or before, the $(N+M$-1)-th trial, there must have been at least N successes in $N+M$-1 trials. Thus also

$$u(M,N)=P(S_{N+M-1}\geq N)=p^N q^{M-1}\sum_{m=0}^{M-1}\binom{N+M-1}{N+m}\left(\frac{p}{q}\right)^m. \tag{5.6.9}$$

When M and N are large the sums become awkward to evaluate numerically and it is then that the Central Limit Theorem furnishes a convenient approximation. As it happens it also provides insight.

Let W be a random variable with finite mean and standard deviation E(W), SD(W). Let S_n be the sum of the n independently and identically distributed random variables W_m. The Central Limit Theorem states that as $n->\infty$,

$$S_n^* = \frac{\{S_n - E(S_n)\}}{SD(S_n)}$$

$$\text{where } S_n = \sum_{m=1}^{n} W_m, \quad , E(S_n) = nE(W),$$

$$SD(S_n) = (n)^{\frac{1}{2}} SD(W)$$

tends to be distributed so that

$$P(\alpha < S_n^* < \beta) \sim \frac{1}{\sqrt{(2\pi)}} \int_\alpha^\beta e^{\frac{-u^2}{2}} du,$$

the ~ sign meaning that the ratio of the two sides -> 1.

Applying this to the random variable T_N, whose mean is known to be N/p and variance Nq/p^2 gives, as $N \to \infty$,

$$P\left(\alpha < \frac{p(T_N - \frac{N}{p})}{\sqrt{(Nq)}} < \beta\right) \sim \frac{1}{\sqrt{(2\pi)}} \int_\alpha^\beta e^{\frac{-u^2}{2}} du,$$

and hence

$$\text{as } N \to \infty$$

$$u(M,N) \sim \frac{1}{\sqrt{(2\pi)}} \int_{-\sqrt{(Nq)}}^{\frac{pM - Nq - p}{\sqrt{(Nq)}}} e^{\frac{-u^2}{2}} du. \qquad (5.6.10)$$

From this it is obvious that if $pM\text{-}Nq=0$, $u(M,N) \sim ½$.

This applies directly to the ancient warfare model since the probability of success in an engagement turns out to be independent of the force sizes when the engagement takes place. For, to conform with the deterministic model, p_{xy} must be proportional to μxy, and q_{xy} to λxy. Since their sum is 1 it follows that $p_{xy} = \mu xy/(\mu xy + \lambda xy) = 1/(1+\rho)$, $q_{xy} = \rho/(1+\rho)$, where $\rho = \lambda/\mu$. The foregoing theory then applies exactly, $pM\text{-}Nq = (\mu M\text{-}\lambda N)/(\lambda + \mu)$, and if $\mu M\text{-}\lambda N = 0$ the probability of victory tends to 1/2. This seems a reasonable probabilistic interpretation of the deterministic criterion for a stalemate.

5.6.4 The mixed model expressed by (5.6.3) leads to a similar Central Limit type approximation for $u(M,N)$. It is shown in [2] that as $N\to\infty$,

$$u(M,N) \sim \frac{1}{\sqrt{(2\pi)}} \int_A^B e^{\frac{-u^2}{2}}\, du,$$

where

$$\sigma B = M - \frac{1}{2}\rho N(N+1) - 1, \quad \sigma A = -\frac{1}{2}\rho N(N+1)$$

and

$$\sigma^2 = \frac{\rho^2}{6} N(N+1)(2N+1), \quad \rho = \frac{\alpha}{\mu}.$$

This throws the invariant (5.6.3a) into prominence in a probabilistic context. Exact formulae for $u(M,N)$ in the form of sums are available but, like (5.6.9), suggest nothing about parametric relationships. The single trial probability of success in this case depends on one of the two combatants' force size, not both. The Central Limit Theorem has to be extended to sums of independent but not identically distributed random variables, but this is legitimate.

5.6.5 This brings us to our focus. What about the square-law model expressed by (5.6.2)? A problem now is that p_{xy} depends on *both* x and y, namely,

$$p_{xy} = \frac{\beta x}{\beta x + \alpha y}.$$

R.H. Brown [3] made an onslaught on $u(M,N)$ using the formula

$$u(M,N) = p_{MN} u(M,N-1) + q_{MN} u(M-1,N),$$

with

$$u(0,N) = 0, \quad u(M,0) = 1. \tag{5.6.11}$$

This is the Chapman-Kolmogorov backward equation. The boundary values are imposed for consistency. Brown (a student of B. Koopman) obtained the exact result

$$u(M,N)=\frac{(-1)^M}{\rho^N}\sum_{t=1}^{M}\frac{(-1)^t\Gamma(1+t/\rho)t^{M+N}}{\Gamma(N+1+t/\rho)t!(M-t)!},\qquad (\rho=\lambda/\mu).$$

The argument used was, however, analytic rather than probabilistic and it must be said that it is a result to be admired, rather than useful, deserving a luminous spot in the Museum of Mathematical Accomplishment. Not only does it give no obvious guidance concerning which of X and Y has the advantage, but it is also extremely unsuitable as a computational vehicle. Try it and see!

It must also be said in fairness that R.H. Brown recognised these criticisms and offered an approximation that now looks familiar, though he did not get it by probability argument.

5.6.6 In the remainder of this discussion we investigate whether the invariant $M-N\surd(\alpha/\beta)$ plays the same kind of rôle in the stochastic theory as its counterparts do in the linear and mixed model frameworks. The approach is numerical. To begin with we calculate $u(M,N)$ exactly from (5.6.11) for a zero value of the invariant. Table 5.6.2a gives the results.

Table 5.6.2a Probability u(M,N) that X is the Victor

$M-N\surd(\alpha/\beta) = 0,\ \alpha/\beta = 0.4444$

M	N	$u(M,N)$
10	15	0.5133
20	30	0.5097
30	45	0.5080
40	60	0.5070
50	75	0.5062

It is seen that as M,N increase while preserving the value of M/N, $u(M,N)$ approaches 0.5.

Next, with the same value of α/β we investigate $u(M,N)$ when $M^2-N^2(\alpha/\beta)=K$ for both positive and negative *fixed* K.

The results are given in Table 5.6.2b.

Table 5.6.2b Probability $u(M,N)$ that X is the victor

$$M^2\text{-}N^2(\alpha/\beta)=K,\ \alpha/\beta=0.4444$$

$K=50$			$K=-50$		
M	N	$u(M,N)$	M	N	$u(M,N)$
10	11	0.8937	10	18	0.2196
20	28	0.6665	20	32	0.3563
30	44	0.5730	30	46	0.4435
40	59	0.5633	40	61	0.4510
50	74	0.5566	50	76	0.4561

The figures suggest that:

(i) as $N\text{-}>\infty$, while $M^2\text{-}N^2(\alpha/\beta)$ remains fixed, $u(M,N)\text{-}>0.5$;

(ii)when $K>0$, X has the higher probability of victory but when $K<0$ the reverse is true.

5.6.7 5.6.6 is based on exact calculation of $u(M,N)$ from a deterministic formula. What happens if we simulate 100 battles using the same value of $\alpha/\beta=0.4444$? The combined results follow.

Table 5.6.3 Results of 100 simulated Battles fought to termination

M	N	S	T	$S/(S+T)$
10	15	526	474	0.526
10	11	896	104	0.896
10	18	233	767	0.233

Notes: 1) S=total number of X-victories, T=total number of Y=victories.

2) $S/(S+T)$ estimates probability of an X-victory.

The results are consistent with Table 5.6.2. In this computer age the interesting possibility suggests itself that an enterprising general who had solved the inverse problem might want to go on and find his probability of success in a later battle by computational methods.

5.6.8 Finally let us examine the possibility of a normal approximation to $u(M,N)$ and its implications when $M\text{-}Nr$ figures in the upper limit ($r=\sqrt{(\alpha/\beta)}$. We recall from **5.6.4** that the crucial random variable is T_N, the number of engagements to the N-th success, and that

$$u(M,N)=P(N\le T_N\le M+N-1)$$

exactly. The normalised variable T_N^* is

$$T_N^*=\frac{T_N-m}{s},$$

where m and s are a mean and a standard deviation which have to be interpreted here. With this

$$u(M,N)=P(A\le T_N^*\le B),$$

where

$$A=\frac{N-m}{s},\quad B=\frac{M+N-1-m}{s}.$$

The crude Central Limit approximation is

$$u(M,N)\sim\frac{1}{\sqrt{(2\pi)}}\int_A^B e^{-u^2/2}du. \qquad (5.6.12a)$$

If by analogy with the constant probability p of success case we put $m=N/p$, $s=\surd(Nq)/p$, with the purpose of finding the "equivalent" p we get

$$A=-\sqrt{(Nq)},\qquad B=\frac{M-\frac{Nq}{p}-1}{\frac{\sqrt{(Nq)}}{p}}.$$

If now we put $q/p=r(=\surd(\alpha/\beta))$ we find $p=1/(1+r)$, $q=r/(1+r)$.

With $\alpha/\beta=0.4444$ we have $r=0.6667$ and $p=0.6$, $q=0.4$. Now we know that probability of success in an engagement in this case varies with force size and therefore we have to conclude that for the postulated normal form to hold *the ratio of force sizes remains nearly constant throughout most of the battle*. Another way of saying this is that the two dimensional random walk executed by the force size two-dimensional random variable (m,n) deviates "on average" not much from a fixed path.

The following Table 5.6.4 supports this conclusion. It is based on two random battles with fairly large initial force strengths $M=100, N=150$, and $r=0.6667$. The force sizes, ratio m/n and $1/(1+r)$ are given for decrements of 10 in m.

Table 5.6.4 Two instances of a large battle

Battle 1				Battle 2			
m	n	$r=m/n$	$1/(1+r)$	m	n	$r=m/n$	$1/(1+r)$
100	150	0.6667	0.6	100	150	0.6667	0.6
90	144	0.6250	0.6154	90	135	0.6667	0.6
80	135	0.5926	0.6279	80	119	0.6723	0.5980
70	121	0.5785	0.6335	70	101	0.6931	0.5906
60	103	0.5825	0.6319	60	86	0.6977	0.5890
50	83	0.6024	0.6241	50	80	0.625	0.6154
40	70	0.5714	0.6364	40	65	0.6154	0.6190
30	44	0.6818	0.5946	30	54	0.5556	0.6428
20	14	1.4286	0.4118	20	45	0.4444	0.6923
17	0	FINISH		10	35	0.2857	0.7778
				0	29	FINISH	

It is seen that the values of r remain nearly constant for a large part of the battles. The mean values of $1/(1+r)$ taken over the whole battle are, respectively, 0.6084 and 0.5961, to compare with the initial value 0.6. $(1+r)^{-1}$ is, of course, the current probability of success.

Finally, Table 5.6.5 shows in similar fashion an instance of each of two battles with the same $r=0.6667$ but initial values (100,110) and (100,180). X has the advantage in the former, and Y in the latter.

Table 5.6.5 Two more Battles with unequally matched Adversaries

m	n	$m/(m+r^2n)$	m	n	$m/(m+r^2n)$
100	110	0.6716	100	180	0.5556
90	88	0.6971	90	173	0.5393
80	77	0.7004	80	161	0.5279
70	39	0.8015	70	150	0.5122
60	7	0.9507	60	142	0.4874
60	0	FINISH	50	136	0.4527
			40	130	0.4091
			30	123	0.3543
			20	119	0.2744
			10	117	0.1613
			0	115	FINISH

The third column in each sub-table is the probability of success at that point in the battle and will be seen to wax and wane more than in Table 5.6.4, not surprising since as time progresses the advantage of the stronger combatant dominates the adversary.

The reader may wish to investigate further so here is another topic: What sort of approximation do the normal integrals give to $u(M,N)$?

References

1.Lanchester, F.W.(1917) *Aircraft in warfare:the dawn of the fourth arm*. Constable.

2.Conolly, B.W. (1981) *Techniques in operational research*. Ellis Horwood.

3.Brown,R.H. (1963) Theory of combat:the probability of winning. Opns. Res.,**11**,418-425.

5.7 BESSEL FUNCTIONS AND DETECTION

5.7.1 This is an exercise which, on the one hand, finds the solution of a practical problem in terms of known and exhaustively studied mathematical functions, and, on the other hand, demonstrates the uselessness of the representation for practical purposes. Thus, given a Personal Computer and some instruction in Numerical Methods, the probabilities sought are least painfully obtained by numerical integration. The scene is illustrated by the diagram.

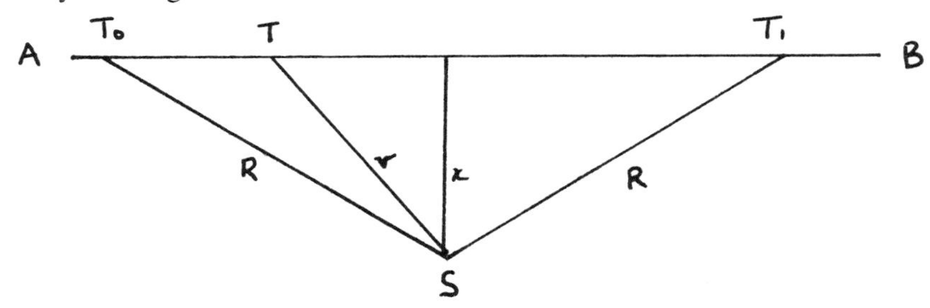

A target moves with speed V on a straight track AB and passes within detection range of a sensor, that is, a detection device, located at a point S at perpendicular distance x from AB. Unless the target passes within range R detection is not possible, and while it is within that range, at distance r, say, the probability that, if exposed over a small time interval $(t,t+h)$ it is detected, is $\lambda(r)h$. With the geometry shown the target becomes first detectable at T_0, and passes out of range at the point T_1. After passage of time t, measured from when the target arrives at T_0, let the range ST be r. Since, as t increases, the distance to the closest point of approach is decreasing, the kinematics give

$$Vdt=-d\{(r^2-x^2)^{1/2}\},$$

and the probability $q(t)$ that the target has **not** been detected by time t is thus

$$q(t+h)=q(t)\{1-\lambda(r)h\}, \tag{5.7.1}$$

since, if not detected by time $t+h$ it can not have been detected by time t, and then must not be detected during $(t,t+h)$, an independence assumption being embedded. It follows by the usual process of expanding the left hand side at t, cancelling $q(t)$ on either side, dividing by h, and then allowing h to vanish, that q satisfies the differential equation

$$\frac{1}{q}\frac{dq}{dt}=-\lambda(r),$$

whence

$$q(t)=\exp\{-\int_0^t \lambda(u)du\}, \tag{5.7.2}$$

since $q(0)=1$ (no detection by time $t=0$). If the detection law is the so-called **definite** one, $\lambda(r)=\lambda$, a positive constant, for $0\leq r\leq R$, and otherwise zero; and so

$$q(t)=e^{-\lambda t}$$

where (5.7.3)

$$Vt=T_0T=(R^2-x^2)^{1/2}-(r^2-x^2)^{1/2}.$$

Thus, the probability $Q(R,x)$ that the target is not detected when travelling from T_0 to T_1 by the sensor S located at closest distance x to its track, is, by symmetry,

$$Q(R,x)=\exp\{\frac{-2\lambda(R^2-x^2)^{1/2}}{V}\}. \tag{5.7.4}$$

For a sufficiently wide area, and given that x is uniformly distributed over $(-R,R)$, so that we are discussing detection by a sensor known to be within detection range, we get

$$Q(R)=\frac{1}{R}\int_0^R \exp\{\frac{-2\lambda(R^2-x^2)^{1/2}}{V}\}dx \tag{5.7.5}$$

for the probability that the target remains undetected throughout its passage, even though it is within range of the sensor.

$Q(R)$ is fundamental in many military and search and rescue applications, but here the emphasis is on its existence as a mathematically interesting function in its own right. Putting $x=R\cos\theta$ enables us to rewrite it in the form

$$Q(R)=\int_0^{\pi/2} e^{-2\lambda R\sin\theta/V}\sin\theta d\theta, \qquad (5.7.6)$$

and $\sin\theta$ can, of course, be replaced by $\cos\theta$. The family resemblance to (5.7.23) in the Notes will not be missed.

It is therefore no surprise to find that $Q(R)$ satisfies a form of Bessel's differential equation. It is of order $\nu=1$, but inhomogeneous.

5.7.2 Differential Equation and Bessel Function Solution

It is convenient to define the functions

$$f(z)=\int_0^{\pi/2} e^{-z\sin\theta}\sin\theta d\theta, \quad g(z)=\int_0^{\pi/2} e^{-z\sin\theta}d\theta. \qquad (5.7.7)$$

Then

$$Q(R)=f(2\lambda R/V),$$

and

$$g'(z) = -f(z), \qquad (5.7.8)$$

where primes denote differentiation with respect to argument (which may sometimes be omitted).

We show first that f(z) satisfies the differential equation

$$f''(z)+\frac{1}{z}f'(z)-(1+\frac{1}{z^2}f(z))=-\frac{1}{z^2} \qquad (5.7.9)$$

$$\textit{with } f(0)=1,\ f(\infty)=0,\ f'(0)=-\pi/4.$$

The first two boundary conditions are obtained easily from (5.7.7). Also, by differentiation,

$$f'(0)=-\int_0^{\pi/2}\sin^2\theta d\theta=-\frac{\pi}{4}.$$

Next,

$$g''(z) = \int_0^{\pi/2} e^{-z\sin\theta}(1-\cos^2\theta)d\theta$$

$$= g(z) - \int_0^{\pi/2} e^{-z\sin\theta}\cos^2\theta d\theta.$$

Also, integration by parts gives

$$f(z) = 1 - z\int_0^{\pi/2} e^{-z\sin\theta}\cos^2\theta d\theta.$$

Thus

$$g''(z) = g(z) - \frac{1-f(z)}{z},$$

and a further differentiation gives

$$g'''(z) = g'(z) + \frac{1}{z^2} + \frac{f'(z)}{z} - \frac{f(z)}{z^2},$$

from which (5.7.9) follows.

Now

$$f''(z) + \frac{1}{z}f'(z) - (1 + \frac{1}{z^2})f(z) = 0$$

is the form of Bessel's differential equation satisfied by the modified functions $I_1(z)$ and $K_1(z)$ (See Notes, (5.7.20)). Thus, the general solution of (5.7.9) can be found by the variation of parameters method from the postulated solution

$$f(z) = A_1(z)\, I_1(z) + A_2(z)\, K_1(z) \tag{5.7.10}$$

where A_1 and A_2 are functions of z to be determined by the standard method sketched below.

For the record, the solution is

$$f(z) = -\frac{\pi}{2} I_1(z) - I_1(z)\int_0^z \frac{K_1(u)}{u} du + K_1(z)\int_0^z \frac{I_1(u)}{u} du. \qquad (5.7.11)$$

To obtain this from (5.7.10) we impose the condition on A_1 and A_2 that

$$A_1'(z)I_1(z) + A_2'(z)K_1(z) = 0.$$

Substitution of the derivatives of f into (5.7.10) gives the additional condition

$$A_1'(z)I_1'(z) + A_2'(z)K_1'(z) + z^{-2} = 0.$$

The Wronskian is known to be $-1/z$, and accordingly

$$A_1'(z) = -K_1(z)/z \quad A_2'(z) = I_1(z)/z$$

and therefore

$$f(z) = EI_1(z) + FK_1(z) - I_1(z)\int_0^z \frac{K_1(u)}{u} du + K_1(z)\int_0^z \frac{I_1(u)}{u} du, \qquad (5.7.12)$$

where E and F are integration constants to be chosen to satisfy the boundary conditions. To satisfy $f(0)=1$ it is found from the series expansion in the Notes that $F=0$. Differentiation gives $f'(0)=E/2$ so that $E=-\pi/2$, and hence the solution (5.7.11).

5.7.3 Numerical Values

As the problem is a practical one, it is natural to ask for numerical values of the probabilities. Aesthetically pleasing as the representation (5.7.11) may appear, it is fraught with serious difficulties. For example:

(i) $I_1(z)$ increases exponentially with z;

(ii) $K_1(z)$ has a singularity at $z=0$ and, although it has to cancel in (5.7.11), this

has explicitly to be organized before starting on (5.7.11) numerically.
Alternatively, the differential equation (5.7.9) could be solved numerically for *f(z)* or, equally plausibly, a suitable numerical integration procedure might be applied directly to the integral definition (5.7.7). The fact of the matter is that *f(z)* is a smooth, well-behaved function with a horrid representation in terms of beautiful and well-loved functions!

A short table of *f(z)* follows. It was calculated from the integral by a nice implementation of Simpson's Rule based on an Algol procedure due to Naor. The program is given in the Notes, **5.7.5.**

Table 5.7.1 Values of *f(z)*

z	*f(z)*	*z*	*f(z)*
0.0	1.00000	6.5	0.20210
0.5	.77774	7.0	.19316
1.0	.62354	7.5	.18530
1.5	.51468	8.0	.17832
2.0	.43633	8.5	.17206
2.5	.37877	9.0	.16641
3.0	.33555	9.5	.16128
3.5	.30239	10.0	.15659
4.0	.27638	15.0	.12469
4.5	.25555	20.0	0.10653
5.0	.23853		
5.5	.22436		
6.0	.21239		

5.7.4 Series Expansions of f(z)

The following are given for completeness rather than for their usefulness.

$$f(z)=\frac{\pi}{2}\sum_{m\geq 0}\frac{(-\frac{z}{2})^m}{\Gamma(\frac{m}{2}+\frac{1}{2})\Gamma(\frac{m}{2}+\frac{3}{2})}. \qquad (5.7.13)$$

This is an alternating series and, consequently, troublesome to evaluate numerically, the more so as $z->\infty$. It can be split into an odd and an even component, but this is a recipe for numerical disaster. It can also be rearranged as

$$f(z)=-\frac{\pi}{2}I_1(z)+1+$$

$$+\sum_{n\geq 1}\frac{z^{2n}}{\{[2n]^2-1]\}\{[2(n-1)]^2-1\}...\{[2.2]^2-1\}\{[2.1]^2-1\}} \qquad (5.7.14)$$

The asymptotic series, valid as $z->\infty$, is

$$f(z)\sim\frac{1}{z^2}\{1+\sum_{n\geq 1}\frac{1}{v_n}\}, \qquad (5.7.15)$$

where v_n is the expression under the summation in (5.7.14).
These series can be used for evaluation, but straightforward use of the integral is numerically simpler.

5.7.5 Notes

5.7.5.1 The functions associated with the name of F.W. Bessel are defined by Watson [1] as solutions of the linear differential equation

$$x^2\frac{d^2y}{dx^2}+x\frac{dy}{dx}+(x^2-v^2)=0, \qquad (5.7.16)$$

where x and ν are in general unrestricted complex variables. Bessel was led by studies of planetary motion to publish in 1824
an investigation of the behaviour of a certain integral which satisfies such a differential equation. However, this is linked with a type of first order differential equation associated with Count Riccati who in 1724 published a discussion of, effectively, the equation

$$x^m dq=dy+\frac{y^2}{q}dx, \quad where\ q=x^n. \qquad (5.7.17)$$

A more general form, known as Riccati's generalised equation, is

$$\frac{dy}{dx}=P+Qy+Ry^2, \qquad (5.7.18)$$

where P, Q, R are functions of x and P,R are not identically zero.
By the transformation

$$y=-\frac{1}{R}\frac{d(lnu)}{dx}=-\frac{1}{Ru}\frac{du}{dx}, \qquad (5.7.19)$$

(5.7.18) becomes

$$u''-(\frac{R'}{R}+Q)u'+PRu=0, \qquad (5.7.20)$$

where dashes denote differentiation with respect to x. This links Riccati's studies and the investigations stemming from it with Bessel's a century later. But for the probabilist particularly, a piquant flavouring is introduced by the fact that it was James Bernoulli, of the *Ars Conjectandi* and "Bernoulli trials", who, in a letter to Leibniz in 1702 announced essentially the transformation (5.7.19) that links the name of Riccati with Bessel. Indeed, the Bernoulli family, who might be said to be to the mathematics of the time what their contemporaries, the Bach family, were to music, were involved in cognate researches: in particular, Daniel (who discussed the St. Petersburg Paradox), published in 1738 a memoir on the oscillations of a heavy chain under gravity in which a series now recognised as representing a Bessel function appears.

Bessel functions and their relatives crop up in countless physical investigations: for instance, diffusion, heat flow, wave propagation. Students of probability will be well aware of their appearance in applications where exponential distributions play a central rôle. The Von Mises distribution, used for angles, involves a modified Bessel function. The theory has indeed been something of an industry, and although [1] continues to be an invaluable source and reference book it is now in some respects out of date.

5.7.5.2 To complement the text we give some properties of Bessel functions of purely imaginary argument, also called *modified* Bessel functions. These satisfy the differential equation

$$x^2y''+xy'-(x^2+\nu^2)y=0. \qquad (5.7.21)$$

As usual, dashes denote differentiation with respect to the argument x. ν is called the *order*. ([1],para.3.7 (1)). One solution is taken as $I_\nu(x)$, defined by

$$I_\nu(x)=\sum_{m\geq 0}\frac{(x/2)^{\nu+2m}}{m!\Gamma(\nu+m+1)},$$

$$\textit{where } \Gamma(\nu+m+1)=(\nu+m)! \textit{ when } \nu \textit{ is an integer.} \qquad (5.7.22)$$

An integral definition of $I_\nu(x)$ is

$$I_\nu(x)=\frac{(x/2)^\nu}{\Gamma(\frac{1}{2})\Gamma(\nu+\frac{1}{2})}\int_0^\pi e^{-x\cos\theta}\sin^{2\nu}\theta d\theta. \qquad (5.7.23)$$

([1],para.3.71,(9)).

A second independent solution, introduced by Macdonald, and now generally used, is $K_\nu(x)$ defined by

$$K_\nu(x)=\frac{\pi\{I_{-\nu}(x)-I_\nu(x)\}}{2\sin(\nu\pi)}. \qquad (5.7.24)$$

([1],para. 3.7 (6)). It can be shown that the Wronskian of these two independent solutions is

$$I_\nu(x)K_\nu'(x)-I_\nu'(x)K_\nu(x)=-\frac{1}{x}. \qquad (5.7.25)$$

([1], para. 3.7,(19).

K is in some respects more awkward than I and possesses a logarithmic singularity at the origin. In particular, we have,

$$K_0(x) = -\ln(x/2)I_0(x) + \sum_{m\geq 0} \frac{(x/2)^{2m}}{(m!)^2}\psi(m+1), \qquad (5.7.26)$$

and, for integer n ≥ 0,

$$K_n(x) = \frac{1}{2}\sum_{m=0}^{n-1} \frac{(-)^m(n-m-1)!}{m!(x/2)^{n-2m}} + (-)^{n+1}\sum_{m\geq 0} \frac{(x/2)^{n+2m}}{m!(n+m)!}[\ln(x/2) -$$

$$-\frac{1}{2}\psi(m+1) - \frac{1}{2}\psi(n+m+1)], \qquad (5.7.27)$$

([1], para.3.7,(14) and (15)), where $\psi(x)$ is the logarithmic derivative of the gamma function, namely

$$\psi(x) = \frac{d}{dx}\ln\Gamma(x) = \frac{\Gamma'(x)}{\Gamma(x)},$$

with $\psi(1) = -\gamma$, *Euler's constant.*

5.7.5.3 The proposal in the text is to calculate

$$f(z) = \int_0^{\pi/2} e^{-z\sin\theta}\sin\theta d\theta$$

by numerical integration. The procedure is simplified if we put $u = \cos\theta$. Then

$$f(z) = \int_0^1 e^{-z\sqrt{(1-u^2)}}du.$$

"Simpson's Rule" can be expressed in the form

$$\int_{x_0}^{x_2} f(x)dx = \frac{h}{3}(f_0 + 4f_1 + f_2) - \frac{h^5}{90}f^{(4)}(\xi)$$

where $h = x_2 - x_1 = x_1 - x_0$, $f_n = f(x_n)$, $x_0 < \xi < x_1$.

The Pascal implementation which follows tackles integration over the range (0,1) by taking first $h = 1/2$ and using the 3-point Simpson formula. h is then halved and another approximation is obtained by evaluating the sum of the integral approximations over (0,1/2), (1/2,1), and comparing with the previous approximation over (0,1). The procedure continues until successive approximations agree to a prescribed tolerance.

```
program integrate(input, output); uses printer;
var x,y,z,gamma,con:real;m,n,n1:integer;

function f(x:real):real;
begin if x=0 then f:=0 else f:=exp(-n*x-gamma/x +n*ln(x)) end;

function simpson (a,b,delta,v:real):real;
label 1;
var h,j,i:real; n,k:integer;
begin v:=(b-a)*v;n:=1;h:=(b-a)/2;
      x:=a;j:=f(x);x:=b;j:=(j+f(x))*h;
1:    b:=0;
      for k:=1 to n do
      begin x:=(2*k-1)*h+a; b:=b+f(x) end;
      i:=4*h*b+j;
      if abs(v)*delta<abs(i-v) then
         begin v:=i;j:=(i+j)/4;n:=2*n;h:=h/2;goto 1;
         end;
         simpson:=i/3
end;

begin gamma:=4.2;n=5;
         con:=1; for n1:=1 to n do con:=con*n/(n1+1);
         con:=con*exp(gamma);
         y:=simpson(0,15,0.000001,3);y:=y*con;
         writeln(lst,'gamma=',gamma,'n= 'n,'y= ',y:1:5);
         for n:=1 to 5 do writeln(lst);
end.
```

Reference

1. Watson, G.N. (1944) *The theory of Bessel functions*. Cambridge University Press. 2nd edition.

6

Organization and Management

6.1 studies in probability terms the loquaciousness of speech-makers and, more extensively, the effect on planning committee-meeting schedules and hence decision-making. This is a treatment of a serious topic with a hint of levity. 6.2 is a note on a linear-algebraic aspect of personnel management, promotion, transfer and resignation or retirement. 6.3 is an extensive survey of two aspects of queueing theory not commonly encountered in elementary texts and courses: the effect of interaction between service and demand, and, briefly, the negative side of priority treatment. Because queueing theory, although pervasive of almost every aspect of life, is usually treated as highly specialized, enough basics are included to make the essay self-sufficient. 6.4 looks at the elementary mathematics of stock-control with particular reference to uncertain supply.

6.1 SECRETARIAL PREOCCUPATIONS

6.1.1 The Keynote Address

"I am worried", said the Social Secretary, "that the Distinguished Invited Speaker at next week's Annual Conference will overrun the time allotted for the keynote address and that we shall be late for lunch. That upsets the programme for the whole day. It's all right for him since he can, and probably will, leave immediately after he has eaten, claiming an inescapable prior engagement."

"When was the last time that happened?" I asked.

"It must have been five or six years ago", she replied. "But the chef gets nervous and the waiters tend to fidget with the cocktail snacks, not to mention the audience, some of whom in the past have allowed impatience to prevail over politeness. It gives the Society a bad name."

"I don't think you need worry too much", I said, in what I hoped was a comforting voice. "In fact, as I am a betting man, I will wager 29:1 against. That means I will pay you £29 if you are late for lunch, while you will only have to give me £1 if you are unlucky - or, should I say, lucky!".

"That sounds too good to be true" said the Secretary with interest. "But how can you be so confident?".

"The Theory of Probability" I replied, laying my right forefinger along the side of my nose. "It's made fortunes from Monte Carlo to Las Vegas."

"And ruined many hearts and aspirations in the process", added the Secretary with some feeling, sucking her breath in sharply in what sounded suspiciously like a sob.

"But perhaps you would like to make my otherwise depressing day more interesting by revealing your secret. Or would it be too complex for my admittedly microscopic brain?".

"Not in the least" I said ambiguously. "You belittle yourself." And I proceeded to explain as follows.

"I have to make the assumption that all Distinguished Invited Speakers (DIS for short) give speeches whose durations under these circumstances have the same statistical characteristics. This is not unreasonable for was it not Erlang himself, the Father of Operational Research, who showed it to be true for telephone conversations, when the distribution turns out to be exponential? And very convenient that is too, though it is not an essential feature of this argument." Seeing the Secretary's eyes beginning to wander I cleared my throat noisily before continuing.

"I must ask you", I said,"to recall a few of the notions you must have picked up in Statistics I. And as you are a young woman that must have been recent enough for you not to have forgotten completely. Let the distribution of speech lengths T (which are positive random variables) be $F(t)$, $(0<t<\infty)$, with probability density function $f(t)$. Then the probability p_1 that this year's speech will exceed last year's is

$$p_1 = \int_0^\infty f(t_1)dt_1 \int_{t_1}^\infty f(t_2)dt_2$$

$$= \int_0^\infty f(t_1)\{1 - F(t_1)\}dt_1$$

$$= 1 - \frac{1}{2} = \frac{1}{2}.$$

The second integral in the first line is the probability that this year's speech will exceed t_1 which, of course, is 1-$F(t_1)$, and since $f(t_1)$ is the derivative of $F(t_1)$, the final line follows." I wrote this down on a piece of paper and added:

"I would therefore bet evens that this year's DIS would be more garrulous than last year's. But if someone pointed out that what I really want is the probability that last year's speech was too long (exceeded some fixed time T_1) and this year's will be even longer, then I can say that the probability I want is

$$\int_{T_1}^{\infty} f(t_1)dt_1 \int_{t_1}^{\infty} f(t_2)dt_2$$

$$= \int_{T_1}^{\infty} f(t_1)\{1 - F(t_1)\}dt_1 = \frac{\{1 - F(T_1)\}^2}{2},$$

and this is less than 1/2, so my even chance bet is a safe one."

"Now", I continued, "suppose that I want the probability that the speech in 1995 will break the 1993 record. This means that the 1994 speaker was less garrulous than his forerunner in 1993. The probability p_2 is given by

$$p_2 = \int_0^{\infty} f(t_1)dt_1 F(t_1)\{1 - F(t_1)\}$$

$$= \{\frac{1}{2}F^2(t_1) - \frac{1}{3}F^3(t_1)\}_0^{\infty} = \frac{1}{2.3}.$$

Clearly, if n years intervene, the probability is $\{n(n+1)\}^{-1}$, so if $n=5$ the probability is 1/30. Hence my 29:1 bet."

"That sounds convincing" said the Secretary admiringly, "though I am not sure I understood it all. But what I also want to know is the probability that in this record year the DIS's speech will exceed T_1. Can you tell me that, even if I don't remember how long ago it was that the last crashing bore was let loose on the captive audience our Society kindly provides?"

"Yes" I replied encouragingly. If you will read [1] you will see that the answer is

$$\{1 - F(T_1)\}[1 - \ln\{1 - F(T_1)\}].$$

If the distribution were exponential with mean τ, then $F(t)=1-\exp(-t/\tau)$ and the above expression is $(1+t/\tau)\exp(-t/\tau)$. The mean length of the record speech would then be 2τ, that is just twice the average. However, if the average is, say, one hour, or more, we are beginning to talk in terms of record speeches of Fidel Castro proportions, and I suppose that is what you want to avoid."

"It is", replied the by now utterly depressed Secretary. "But what on earth can I do about it?".

"Invite me!", I replied.

6.1.2 Traffic TailBacks

Feller [1] points out in an example that this theory has predictive value also for what he calls "traffic platoons". These may arise by reason of some obstruction, but also when a leading vehicle has a certain speed v, say, and the members of the succession of following vehicles lack the courage to overtake. Thus a platoon, or tailback, builds up, led by the vehicle with the record speed so far (and the longer the tailback, the greater is the deterrent to overtaking). This goes on until some bold spirit "breaks the record". The probability of a tailback of length n is thus $\{n(n+1)\}^{-1}$.

6.1.3 Committee Dynamics

It turned out that my secretary friend had another problem related to loquaciousness. Politicians, academics and union officials, to name some, are notorious talkers. They get plenty of practice and, if you are looking for members of the various species, few locations are richer than important committees. Decisive individuals, on the other hand, tend to taciturnity, brevity, and sometimes lack diplomatic finesse. On the whole, they are reluctant and sometimes unwelcome members of committees. Why this concern with committees? Well, my friend told me that she had recently been appointed secretary of an important academic committee and that, among other tasks, it fell to her to organise the agenda. But she had found that difficult because the length of the discussion evoked by the various items seemed to bear little relation to their importance. She had, moreover, noticed other disturbing tendencies to illustrate which she showed me records of two recent meetings, obtained by secreting a stop-watch among her secretarial paraphernalia. The results are shown in Figure 6.1.1. Both meetings were scheduled by custom to last for a maximum of three hours (0900-1200) and to have an agenda containing fifteen items, thus allowing an average of twelve minutes for discussion of each item. Figure 6.1.1 plots time horizontally, in minutes, and agenda item numbers on the vertical. The top diagram was obtained from the meeting of a large and, on the whole garrulous and disputatious membership, and it will be seen that the first and the fourth items needed respectively 37 and 48 minutes for decision to be reached. As the end of the meeting approached (after which it was the custom to feed the membership rather well, and at no expense to themselves) it seemed that items were dismissed with increasing rapidity and indeed the one or two minutes actually spent on these final items resulted in the decision to defer them to the next meeting!

On the other hand, the lower diagram relates to a meeting with smaller and less contentious membership. In this case it is seen that the earlier items were dismissed summarily in a few minutes each but that, as time passed, there was a tendency for discussion to be prolonged. It occurred to the observer that perhaps the members - even the Chairman - were worried about what might be said by colleagues who heard that the discussion had been apparently so short. She felt that the phenomenon might be labelled the "better-not-finish-too-early effect", while the contrary phenomenon in the first observations was in fact a "lunch-time effect".

What did I think about this, asked my friend. She did not suppose that mathematics could be of much predictive assistance but she felt it worth while to ask anyway. "On the contrary", I replied. "I have had ample opportunity to reflect on the phenomena you describe when in the past I was forced to serve on various unbelievably tedious committees. I developed a theory which I shall explain right now." This is the gist of what I told her. The Notes contain a summary of the formulae and their derivation. The key to the whole thing is to postulate a plausible decision-making mechanism that reflects human frailty. We are thus crossing the threshold into social mathematics. We introduce a **mean decision-making rate, or intensity,** denoted by $\mu_n(t)$ and defined by

$$\mu_n(t) = \frac{\mu(N-n)}{T-t},$$

where n is the number of items decided by time t after the meeting began, N is the total number of items scheduled, and T the total time allocated to the session. μ is a positive constant which we label the **index of decisiveness** in the sense that the larger μ is, the more rapid the decision-making. It depends both on committee membership measured in size and garrulousness. In fact, the more talkative the members, the smaller is μ. The same goes for committee size - the larger a committee, the smaller the index of decisiveness. The postulated mechanism incorporates the following two features:

(a) the factor N-n implies pressure on the decision makers when the number of items still to be dealt with is large; this accords with the well-known human tendency to try to deal quickly with a long list of distasteful tasks;

(b) the factor T-t in the denominator also tends to accelerate the speed of decision making as the scheduled time for the end of the meeting approaches. We have called this the **lunchtime effect**, but exigencies other than hunger might share the responsibility.

Statistically inclined readers will notice that the more the items on the agenda ($N \to \infty$), the larger must T be to accommodate them (in a sensible world, that is) and we can postulate, if we like, that T and N tend to infinity in such a way that their ratio $N/T = \lambda$. In these circumstances $\mu_n(t) \to \lambda\mu$, and the index becomes constant. As we shall see, the statistical consequence is that very long committee meetings become Poisson processes. But we are rather concerned with finite N and T.

To construct a realistic mechanism for predictive purposes we need to think in terms of random variables. The obvious candidates are:

(i) $\nu(t)$=the number of decisions taken by time t.
This is a positive integer with the extreme values $\nu(0)=0$, and $\nu(T)=N$.

(ii) The time T_n to the n-th decision ($T_0=0$). This is a positive real-valued random variable. In the nature of our friend's observations we shall also be interested in

$$\tau_n = T_n - T_{n-1},$$

the time to make the n-th decision. Plainly,

$$T_n = \tau_1 + \ldots + \tau_n.$$

Figure 6.1.1 Records of Two Committee Meetings

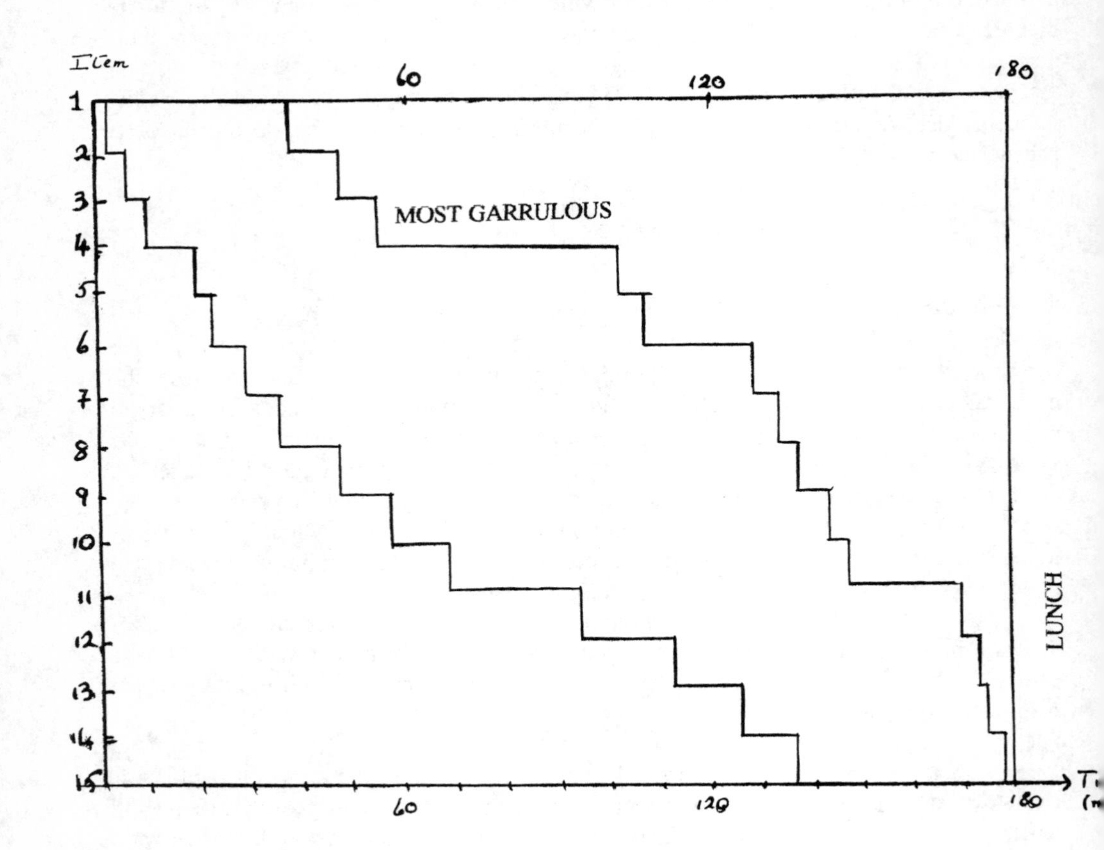

It happens that $\nu(t)$ is binomially distributed with probability of success $\alpha=(1-t/T)^{\mu}$. Thus

$$p_n(t)=\binom{N}{n}\alpha^{N-n}\beta^n, \quad (\beta=1-\alpha).$$

Also, the probability density function $h_n(t)$ of the time T_n to the n-th decision is given by

$$h_n(t)=\frac{\mu}{T}\frac{N!}{(n-1)!(N-n)!}\alpha^{(N-n+1)-\frac{1}{\mu}}\beta^{n-1}, \quad (n=1,2,...,N).$$

We then find that

$$E(T_1)=E(\tau_1)=\frac{T}{N\mu+1}$$

$$E(T_n)=T[1-\frac{N(N-1)...(N-n+1)}{(N+\frac{1}{\mu})(N-1+\frac{1}{\mu})...(N-n+\frac{1}{\mu}+1)}],$$

and hence

$$E(\tau_n)=E(T_n)-E(T_n-1)=\frac{T}{\mu}\frac{N(N-1)...(N-n+2)}{(N+\frac{1}{\mu})...(N-n+\frac{1}{\mu}+1)}, \quad (2\leq n\leq N).$$

Let us look in Figure 6.1.2 at what these last formulae predict for mean decision time to the n-th item, T_n, for values of μ corresponding to lower ($\mu=0.5$), medium ($\mu=1$), and higher ($\mu=2$) mean decisiveness rates. The format is as for Figure 6.1.1. Again the scheduled meeting duration $T=180$ minutes, and the total number N of agenda items is 15. It is seen that the tendencies referred to in Figure 6.1.1 are reinforced. As time passes, the more decisive ($\mu=2$) committee tends to slow down, while the less decisive one ($\mu=0.5$) accelerates. Interestingly, when $\mu=1$, the mean decision time of each item is equal: thus $\mu=1$ represents a kind of behaviour norm, a secretary's ideal, with no fuss, no nonsense and a uniform decision-making rate. This suggests the need for a thorough study of the relationship with μ of committee size and garrulousness. My secretary friend looked dismayed, but I continued. "It is unwise", I said, "to repose utter trust in mean values. What we must do next is to look at some probabilities. Then we can bet on events like 'at least half the agenda has been completed by half past eleven', have some fun and maybe make some money. Meetings will resemble horse-races and people will fight to get into the public gallery to watch the spectacle."

Figure 6.1.2 Mean Times to make Decisions

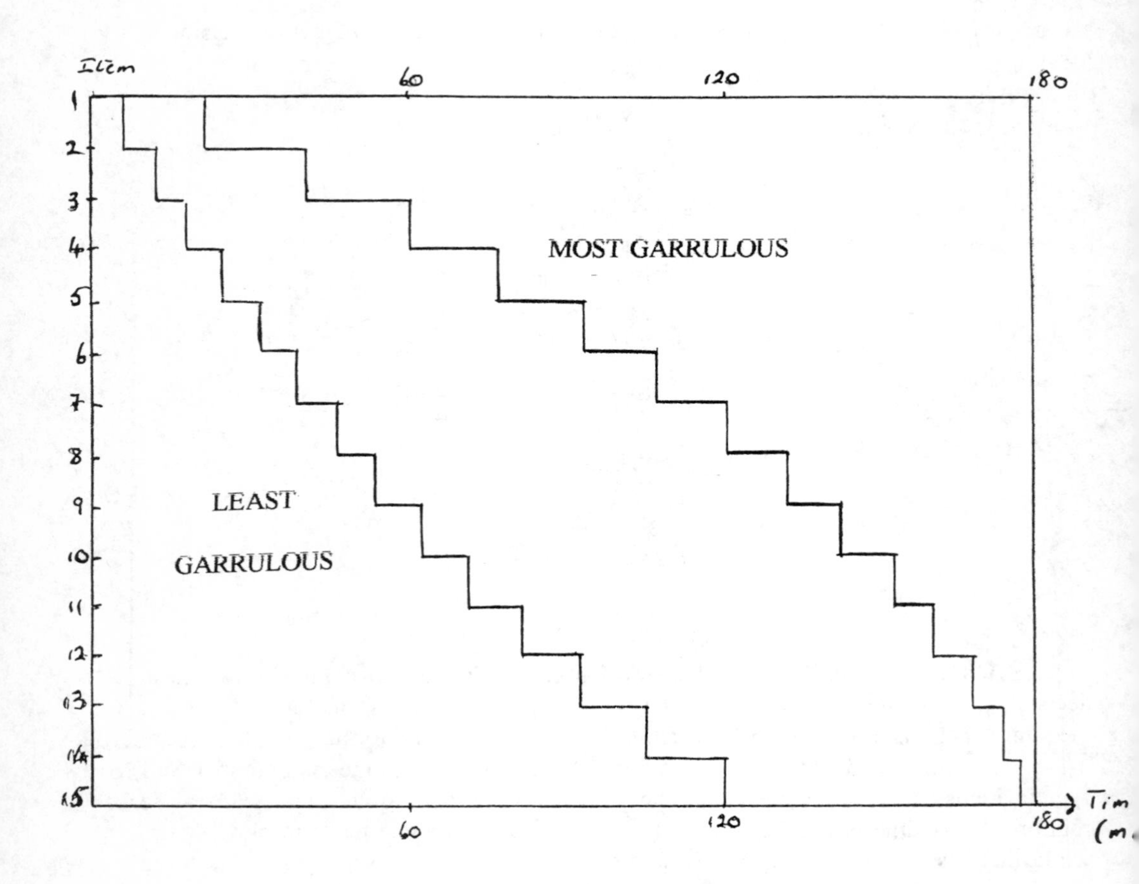

In this spirit I have prepared Figure 6.1.3 which plots the probability that the agenda is more than half completed by time t shown on the horizontal axis. $t=90$ is the half way mark and you see that, whereas it is virtually certain that the more decisive committee ($\mu=2$) is more than half way to the finishing post, that is a highly improbable state of affairs if the committee is an indecisive one ($\mu=0.5$). You may be surprised to notice that when $\mu=1$ there is an even chance that the agenda is half finished by half time."

Figure 6.1.3 Probability that Agenda is at least half finished by Time t

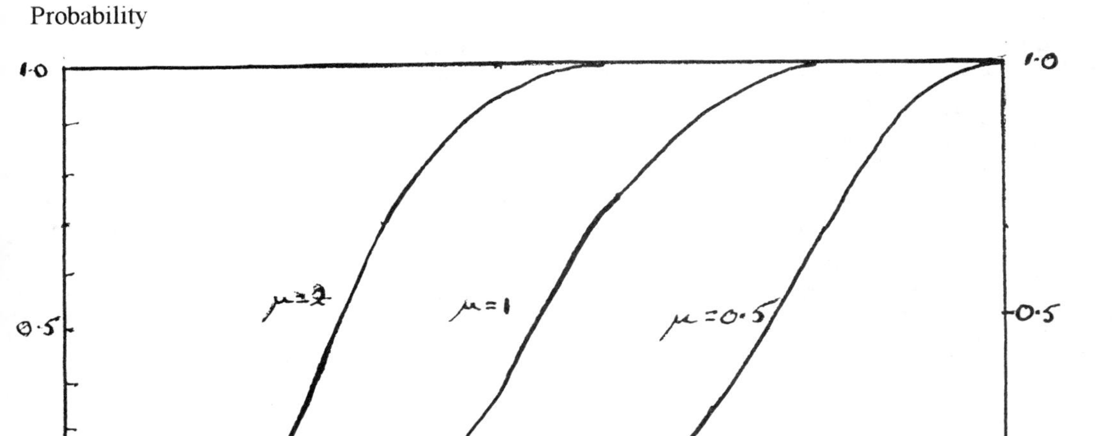

I went on to point out that when $\mu = 1$ the decision times of each item are statistically identical - that is they have the same distribution - but that is not the case when $\mu \neq 1$. Obtaining the probability density function $g_n(\tau)$ in this last case is an ugly business the flavour of which can be tasted in the Notes.

"It is also shown in the Notes how to simulate a committee meeting on a digital computer", I went on, but was interrupted by my friend who asked if that meant that committees could be replaced by robots, or randomising machines, something like an amusement arcade. "Now you are being flippant", I said severely, "although the prospect stimulates diverting thoughts. However, there are lessons to be learned from the theory. For example, you must find out how more or less garrulous your members are, and you can do that by, on some pretext, asking individuals their birth-dates, and if it takes more than ten minutes to get the information, you conclude that you are dealing with a person with high garrulousness coefficient. Then, if your committee is highly garrulous, you must be sure to place important items early in the agenda. The obvious reason is to ensure adequate discussion: even if the early item were unimportant, like six new teaspoons for the manager's office, it would still evoke the same florid oratory, and then the important item, if placed at the end, would almost certainly be deferred to the next meeting, raising thus the possibility that a decision is never reached at all. Conversely, important items should be placed later on the agenda of committees with laconic membership so that the 'better-not-finish-too-early effect' can be relied on to operate.

Finally, it is not a bad idea to ascertain the equivalent hourly rate of pay of committee members and to display prominently during the meeting the cost so far. Members can then work out for themselves the price of their decision-making."

6.1.4 Notes

6.1.4.1 The results in **6.1.1** are obtained as follows. First suppose that the previous record occurred n years ago and had value x. The joint probability and probability density function of this, and that the new record has value y is

$$f(x)dx\ F^{n-1}(x)\ f(y)\ dy,$$

where f is the unspecified probability density function of the positive random variable X with value x, and F its distribution function. Summing over all possible values of x $(0<x<y)$ gives

$$f(y)dy \int_0^y F^{n-1}(x) f(x) dx = \frac{f(y) F^n(y)}{n}.$$

Thus, the probability that the record value 'n years later' lies in the range $(0, T_1)$ is $F^{n+1}(T_1)/\{n(n+1)\}$. By putting $T_1 = \infty$ we get the probability $\{n(n+1)\}^{-1}$ that it has some value.

The distribution of T_1, whatever n, is obtained by summing over $n \geq 1$. This gives $F(T_1) + \{1 - F(T_1)\}\ln\{1 - F(T_1)\}$, and the probability that a record value exceeds T_1 is

$$\{1\text{-}F(T_1)\}[1\text{-}\ln\{1\text{-}F(T_1)\}].$$

If $F(T)=1\text{-}\exp(-T/\tau)$, this probability is $\exp(-T/\tau)(1+T/\tau)$, and its value when $T=2\tau$, the mean, is 0.41, so indeed speeches of gargantuan proportions are not unlikely.

6.1.4.2 The following is a summary of the essentials of the committee dynamics analysis.

(i) **Decision Intensity**

This is

$$\mu_n(t)=\frac{\mu(N-n)}{T-t},$$

where μ is the **index of decisiveness** in the sense that, the larger μ, the more rapidly decisions are reached. It depends inversely on the committee size and on the garrulousness of the members. Notice that the law entails a slowing of the decision processes as the end of the agenda is approached (as $n\text{-}>N$), and acceleration as the scheduled time for completion is neared (as $t\text{-}>T$).

$N=$ number of items on the agenda; $T=$total time scheduled.

(ii) $p_n(t)=$P[**number of decisions made by time** $t=n$], $(n=0,1,\ldots,N)$.

Let T_n be time to n-th decision and τ_n=time from $(n$-1)-th to n-th

decision. Then

$$T_n=\tau_1+\ldots+\tau_n.$$

$$p_0(t)=P(T_1>t)$$

$$p_1(t)=P(T_1<t,T_2=\tau_1+\tau_2>t)$$

$$p_2(t)=P(T_2<t,T_3=\tau_1+\tau_2+\tau_3>t)$$

....

$$p_n(t)=P(T_n<t,T_{n+1}>t)$$

....

$$p_N(t)=P(T_N<t).$$

Also

$$p_n(t)=\binom{N}{n}\alpha^{N-n}\beta^n,\ \ \alpha=(1-t/T)^{\mu},\ \ \beta=1-\alpha.$$

Note the check that the sum of the probabilities is 1. Note also that if both N and $T\to\infty$ in such a way that $N=\lambda T$, $p_n(t)\to$ Poisson with mean rate $\lambda\mu$. The form of $p_n(t)$ is obtained by the "usual" differential difference equation technique from the Chapman-Kolmogorov forward equations, leading to

$$\dot{p}_0=\frac{-N\mu}{T-t}p_0$$

$$\dot{p}_1=\frac{N\mu}{T-t}p_0-\frac{(N-1)\mu}{T-t}p_1$$

....

$$\dot{p}_n=\frac{(N-n+1)\mu}{T-t}p_{n-1}-\frac{(N-n)\mu}{T-t}p_n$$

$$\dot{p}_N=\frac{\mu}{T-t}p_{N-1}.$$

The dependence on t has been omitted.

(iii) **Probability density function $h_n(t)$ of $T_n=\tau_1+\ldots+\tau_n$**
Since $p_0(t)=P(\tau_1>t)$, $h_1(t)=-dp_0/dt=[N\mu/(T-t)]p_0$.

Also

$$p_0(t)+p_1(t)=P(\tau_1>t)+P(\tau_1<t,\tau_1+\tau_2>t)$$

$$=P(\tau_1+\tau_2>t),\ so$$

$$h_2(t)=-[\dot{p}_0(t)+\dot{p}_1(t)]=\frac{(N-1)\mu}{T-t}p_1$$

and so on.This gives

$$h_n(t)=\frac{\mu}{T}\frac{N!}{(n-1)!(N-n)!}(1-\frac{t}{T})^{(N-n+1)\mu-1}\{1-(1-\frac{t}{T})^{\mu}\}^{n-1}$$

$$(n=1,..,N).$$

We then get

$$E(T_n) = \int_0^T h_n(t)t\,dt$$

$$= T[1 - \frac{N(N-1)\dots(N-n+1)}{(n+\frac{1}{\mu})(N-1+\frac{1}{\mu})\dots(N-n+\frac{1}{\mu}+1)}],$$

and hence

$$E(\tau_n) = E(T_n) - E(T_{n-1}) = \frac{N(N-1)\dots(N-n+2)T}{(N+\frac{1}{\mu})\dots(N-n+\frac{1}{\mu}+1)\mu}, \quad (n \geq 2).$$

(iv) **Probability densities** $g_n(t)$ **of** τ_n

These are less attractive. $g_1(t)$ is the simplest since it is the same as h_1(t). A general approach is to seek the joint probability and density that $T=t$ and that by time $T+\tau$ the $(n+1)$-st decision has not yet been taken. Denote this by $h_n(t,t+\tau)$. Then over the small time interval $(t,t+\delta)$

$$h_2(t,t+\tau+\delta) = h_2(t,t+\tau)[1 - \frac{\mu(N-2)\delta}{T-t-\tau}]$$

leading to

$$\frac{dh_2(t,t+\tau)}{d\tau} = -\frac{\mu(N-2)}{T-t-\tau} h_2(t,t+\tau)$$

so that

$$h_2(t,t+\tau) = [(T-t-\tau)/(T-t)]^{\mu(N-2)} h_2(t)$$

since clearly when $\tau = 0$,

$h_2(t,t+\tau)$ *must reduce to* $h_2(t)$.

Here is the result for $g_2(\tau)$.

$$g_2(\tau)=\frac{\mu^2N(N-1)}{T}(1-\frac{\tau}{T})^{\mu(N-1)}\int_0^1(1-s)^{\mu(N-1)-1}\{1-(1-\frac{\tau}{T})s\}^{\mu-1}ds.$$

Expansion of the curly parenthesis in powers of s and integration term by term gives the series form

$$g_2(\tau)=\frac{\mu N}{T}(1-\frac{\tau}{T})^{\mu(N-1)}.$$

$$[1-\frac{(1-\frac{\tau}{T})(\mu-1)}{\mu N-\mu+1}+$$

$$+\frac{(1-\frac{\tau}{T})^2(\mu-1)(\mu-2)}{(\mu n-\mu+2)(\mu N-\mu+1)}-...].$$

It is left as an exercise for the reader to finish this, and to show that the sum integrates to unity taken over $0\leq\tau\leq T$. He can then go on to $g_3(\tau)$, and so on.

(v) **Simulation**

Under this heading we show only how to generate pseudorandom intervals τ_n. The basis is an expression for

$$u=P(\tau_{n+1}<t|\tau_1,\tau_2,...,\tau_n).$$

Let $q_n(\tau_1,\tau_1+\tau_2,\ldots,\tau_1+\tau_2+\ldots+\tau_n)$ be the joint probability density function of the times when the first,...,n-th decisions were taken, we want the conditional probability density function

$$\frac{q_{n+1}(\tau_1,\tau_1+\tau_2,...,\tau_1+\tau_2+...\tau_{n+1})}{q_n(\tau_1,\tau_1+\tau_2,...,\tau_1+\tau_2+...+\tau_n)}.$$

It turns out that this conditional density is

$$\mu(N-n)\frac{(T-\tau_1-\tau_2-...-\tau_{n+1})^{\mu(N-n)-1}}{(T-\tau_1-\tau_2-...-\tau_n)^{\mu(N-n)}}.$$

Then

$$P(\tau_{n+1}<\theta\,|\,\tau_1,\tau_2,\ldots,\tau_n)=1-\{1-\frac{\theta}{T-T_n}\}^{\mu(N-n)}.$$

To find a surrogate[1] random τ_{n+1}, given T_n, we generate a pseudo uniform u on (0,1) and then solve

$$(1-\frac{\theta}{T-T_n})=u^{\frac{1}{\mu(N-n)}}.$$

This θ **is** the required pseudorandom τ_{n+1}.

Reference

1. Feller, W. (1966). *An introduction to probability theory and its applications.* Vol.2, Wiley.

6.2 MANPOWER PLANNING

A large agency works world-wide, with three main offices, in Atlanta, Berlin and Canberra. To keep their senior staff efficient and up-to-date, they exchange selected members from time to time, with occasional returns to offices where they have been before. At a meeting of the main board the personnel manager submitted the following statistics:

At the beginning of the year under examination, out of 1000 senior employees, 400 worked in Atlanta, 300 in Berlin, and 300 in Canberra. During the year, 120 of the Atlanta staff remained in Atlanta, 160 were transferred to Berlin, and 40 to Canberra. The remaining 80 left the organization.

Of the Berlin staff, 150 stayed put, 30 went to Atlanta, and 90 to Canberra. 30 left altogether.

Of the Canberra staff, 60 transferred to Atlanta, 30 to Berlin, 120 remained in Canberra, while 90 left.

The personnel manager then asked the board for instructions about the future transfer programme, assuming that the number of leavers will remain constant. The Chairman remarked that if his mental arithmetic was sound there were now:

210 staff members in Atlanta, 340 in Berlin, and 250 in Canberra.

---

[1]The words "surrogate" and "pseudo-" are used because the basic mechanism used for generating "random" orms is in reality deterministic. Tests are made in advance to ensure that the streams of numbers produced fy the important attributes of true uniforms. The surrogate random numbers based on these are thus also udo".

He suggested that the loss of 200 should be replaced by recruiting the same number, not necessarily in the location where the wastage had occurred. Regarding a decision on future transfer, he asked for a paper to be prepared about how the numbers would develop if the transfer and loss rates remained the same as they were during the year of scrutiny, namely

	To	A	B	C	Wastage
	A	0.3	0.4	0.1	0.2
From	B	0.1	0.5	0.3	0.1
	C	0.2	0.1	0.4	0.3

The answer depends, obviously, on the relative recruitments into the three main offices.

If all recruits were to join Atlanta, the next five years would see the following developments (rounded to integers):

210+200	207+191	204+188	203+188	203+188
340	359	363	363	363
250	243	245	246	246

If the recruits were all to join Berlin, the development would be:

210	167	162	162	162
340+200	379+171	370+173	366+175	365+176
250	283	295	297	297

and for Canberra:

210	187	189	191	192
340	299	276	267	264
250+200	303+211	314+221	316+226	316+228

It appears that in all these cases the distribution pretty much settles down after a few years. If the new recruits were distributed over three locations in the ratios $r_1{:}r_2{:}r_3$, the distributions would tend to

$$\begin{aligned}&391r_1+162r_2+192r_3\\&363r_1+541r_2+264r_3\\&246r_1+297r_2+544r_3.\end{aligned}$$

This knowledge allows management to exercise some influence on future staff numbers and their distribution.

A population which, given the transformation matrix, reproduces itself is called **stationary**. We shall now prove that, whichever ratios r_i we choose, the population tends to stationarity. We shall do even more, namely we shall show how to find the stationary population directly, that is without having to compute the successively emerging populations as was done above. Let us describe in algebraic terms what we have been doing. We started with a population vector, in our case

$$N^{(0)}=(400,300,300)$$

and values p_{ij}, the transition rates from location i to location j. Also we had a wastage rate $w_i=1-p_{i1}-p_{i2}-p_{i3}$. The population vector in the next year is then

$$N_j^{(1)}=N_1^{(0)}p_{ij}+N_2^{(0)}p_{2j}+N_3^{(0)}p_{3j}+r_j\sum_{i=1}^{3}N_i^{(0)}w_i$$

or

$$N_j^{(1)}=N_1^{(0)}s_{1j}+N_2^{(0)}s_{2j}+N_3^{(0)}s_{3j}$$

where

$$s_{ij}=p_{ij}+w_ir_j \quad (i,j=1,2,3).$$

We have multiplied the matrix (s_{ij}) by the vector N_i, and then we considered three cases, those for $r_1=1, r_2=1, r_3=1$.

Let us look more closely at the matrices (s_{ij}) in these three special cases. They are:

$$A_1 = \begin{matrix} 0.5 & 0.2 & 0.5 \\ 0.4 & 0.5 & 0.1 \\ 0.1 & 0.3 & 0.4 \end{matrix} \qquad A_2 = \begin{matrix} 0.3 & 0.1 & 0.2 \\ 0.6 & 0.6 & 0.4 \\ 0.1 & 0.3 & 0.4 \end{matrix}$$

$$A_3 = \begin{matrix} 0.3 & 0.1 & 0.2 \\ 0.4 & 0.5 & 0.1 \\ 0.3 & 0.4 & 0.7 \end{matrix}$$

Observe that the total of each column is unity.

We proceeded by multiplying the population vector of any year by the transition matrix, that is, we computed

$$AN^{(0)},\ A^2N^{(0)} \ldots A^tN^{(0)}.$$

We are interested in the properties of these products. In the following, concepts of matrix theory, explained in the Notes, are used.

The eigenvalues of the matrices A_i are:

	A_1	A_2	A_3
λ_1	1	1	1
λ_2	$0.2+i\surd(0.05)$	$0.15+i\surd(0.0175)$	$0.25+i\surd(0.0175)$
λ_3	$0.2-i\surd(0.05)$	$0.15-i\surd(0.0175)$	$0.25-i\surd(0.0175)$

One of the eigenvalues is equal to unity because the columns add to unity: the other two have smaller moduli. We show in the Notes that under these circumstances the populations tend to an eigenvector corresponding to the unit-valued eigenvalue. In our case the eigenvectors are

27	12	11
25	40	15
17	22	31

and these are indeed proportional to the population structures to which those in our example tend when the total population size is 1000.

We know now that we need not compute successive population structures to find the stationary one. All that has to be done is to compute an eigenvector of the unit eigenvalue. The proof still considers a succession of populations, but even this is not necessary. We can find the stationary population independently of its being a structure to which other structures converge.

What we want to find is such a structure, or such structures, which are reproduced after multiplication by a given transition matrix, and addition of appropriate numbers of recruits which must, of course, be non-negative. These numbers, which will be added into the j-th location, will be called u_j $(j=1,2,3)$.
We have to find non-negative n_i $(i=1,2,3)$ and u_j $(j=1,2,3)$ such that

$$u_j+\sum_{i=1}^{k} s_{ij}n_i=n_j, \quad \textit{with } n_1+n_2+n_3=1.0$$

In our example this means

$$\begin{aligned} u_1 - 0.7n_1 + 0.1n_2 + 0.2n_3 &= 0 \\ u_1 + 0.4n_1 - 0.5n_2 + 0.1n_3 &= 0 \\ u_1 + 0.1n_1 + 0.3n_2 - 0.6n_3 &= 0 \\ n_1 + n_2 + n_3 &= 1000. \end{aligned}$$

This looks like a linear programming problem without a linear form to be optimised. Well, we do not **want** to optimise anything, but to find all solutions of the system with non-negative variables. We know that all solutions lie in the convex hull of solutions with just four variables non-zero, and these solutions are computable by the Simplex Method, provided that we have a method (e.g. computer program) which lists alternative solutions. We obtain the following solutions:

n_1	n_2	n_3	u_1	u_2	u_3
391	363	246	188	0	0
162	541	297	0	176	0
192	264	544	0	0	288

These are precisely the answers obtained above. In each of the four cases, all recruits join just one branch.

6.2.1 Notes

Consider a matrix $A=(a_{ij})$. We want to find values λ such that

$$|A-\lambda I| = \begin{vmatrix} a_{11}-\lambda & a_{12} & \dots & a_{1k} \\ a_{21} & a_{22}-\lambda & \dots & a_{2k} \\ \dots & \dots & & \\ a_{k1} & a_{k2} & \dots & a_{kk}-\lambda \end{vmatrix} = 0.$$

Such values are called eigenvalues of the matrix A. The matrices dealt with here are such that the columns add to unity and therefore one of the eigenvalues is $\lambda_1=1$. Also, it can be shown that in our matrices no other matrix can have a modulus larger than unity. We shall assume all eigenvalues to be different.

The system

$$(A - \lambda I)x = 0$$

has a solution if and only if λ is one of the eigenvalues, and the resulting vector x is called the eigenvector corresponding to the eigenvalue λ used. Eigenvectors are not uniquely defined: if x is an eigenvector, then so is cx, with $c\neq 0$. If all eigenvalues are distinct, then the eigenvectors of the various eigenvalues are linearly independent, and hence any population vector $N=(n_1,\dots,n_k)$ can be expressed as a linear combination of the eigenvectors, choosing one from each eigenvalue. Thus

$$N^{(0)}=c_1v_1+\dots\ +c_kv_k.$$

Now $Av_i=\lambda_i v_i$, and therefore, by induction,

$$A^t v_i=\lambda_i^t v_i$$

for every positive integer t. It follows that

$$N^{(t)}=A^t N^{(0)}=c_0\lambda_1^t v_1+\dots+c_k\lambda_k^t v_k.$$

If $\lambda_1 = 1$, and all other eigenvalues have moduli smaller than unity, then

$$\lim_{t \to \infty}(A^t N^{(0)}) = c_1 v_1,$$

an eigenvector of $A\lambda_1 = 1$. Had we started with $N^{(0)} = v_1$, then the populations would remain unchanged in the successive years: **a stationary population**.

6.3 NEGLECTED QUEUEING THEORY

6.3.0 Introduction

The service industries are making a substantial, inescapable and growing intrusion into modern life. Such an industry supplies something that people want, or think they want, from entertainment to medical care, and at some stage there is usually more demand for the service than can be supplied. At such times the customers may choose to wait, or may have to, for example when they need a new heart or kidney, and they then form a *waiting line*, or *queue*.

Readers probably know how extensive the literature of queueing theory is. After a brief introduction to the ideas and elements necessary to provide a self-contained account, this essay concentrates on two aspects that are usually neglected. The first concerns the advantage that can be gained by cooperation between customers and service, and the second is a comment on the negative side of allocating priority to some customers, important for some, but a possible disaster for others. Non-beginners can skip the introduction, noting merely the notation and terminology and recalling some basic ideas and results that are used. Pertinent references are given. A brief, but more comprehensive, introduction can be found in [7].

6.3.1 Basics

Queues,or **waiting lines**, form when a supply centre, for example a shop, dentist, airport, university, hospital, is unable to satisfy immediately the demands made upon it. It is convenient to use a specialised, though simple, vocabulary.

Supply centre=service, or service point

Individuals or entities requesting service=customers=demand

Those customers who can not receive service immediately, and are prepared to wait, form the **queue**.

The order in which customers are served=**queue discipline**, which may, for example, be first-come, first-served (FCFS),last-come, first served (LCFS), random: or some kind of **priority** may operate for certain categories of customer.

The service may react positively or negatively to pressure of demand, for example work faster, the longer the queue,or, perversely, more slowly; or the demand may modulate its pressure in response to service performance, for example, discouraged, and sometimes encouraged, by long queues. In such circumstances we speak of an **interactive system**.

The number of customers waiting in one or more queues, plus those being served, is called the **system state**. This is a non-negative integer. The total time spent by a

customer waiting for service and then receiving service is called **system time**, and this is a non-negative real number.
When a queueing system operates in such a way that **on average** there is no tendency for the queue to grow indefinitely, we say that the system is in a **steady state**. Such a state implies that the service is more than able to cope with demand. When this is not the case we say that the system is in a state of **congestion**. System state and time are of concern both to the supplier and to the consumer, but of special interest to the supplier are the periods during which the system state is continuously positive. These are called **busy periods.** Their length, measured both in time and in numbers of customers served, is also a direct measure of server fatigue and therefore an important operational characteristic of the system.

The original analysis of queueing systems by Erlang [1] was made in the context of traffic at a telephone exchange, which may now seem an archaic concept, but was in Erlang's time up to the moment. Telephone traffic is characterized by calls of uncertain duration made at irregular time intervals, both capable of adequate description only in probability terms. Our plan, throughout this chapter, will be to assume **exponential** distributions for both **service time** S (i.e. time required to satisfy a customer's demand), and for **inter-arrival intervals , or times,** T, (that is the time intervals separating successive demands for service). This choice is less restrictive than it may seem. Studies by Erlang, and subsequently, have shown the exponential to be a good fit to observation in a wide variety of circumstances. Another important advantage is that exponential distributions allow the analysis to be carried out simply and explicitly. The reader who is interested can pursue the consequences of more sophisticated distributional assumptions in the references given later.

Queueing theory is an appealing topic for its practical interest, its amenability to analysis and its use as a teaching vehicle for students of stochastic processes. It has an enormous literature, but some of this is repetitive and can be ignored.

The purpose in this essay is first to sketch the essential analysis of simple one- and two-server systems driven by exponential mechanisms, and then to discuss the effect of interactive behaviour. Finally, some of the pros and cons will be discussed of systems with simple priority rules. To avoid interrupting the narrative, more than usually tedious details are relegated to the Notes.

6.3.2 One- and two-server exponential systems

6.3.2.1 System state

In the literature, such systems are described briefly by the hieroglyphs **M/M/1** and **M/M/2**. The "M"'s mean "memory-less" or "markovian", that is, exponential. The first "M" refers to the distribution of inter-arrival intervals T, and thus

$$P(T<t)=1-e^{-\lambda t}, \quad (t\geq 0) \qquad (6.3.1)$$

so that λ is the **mean arrival rate** of new customers, and also its reciprocal is the **mean**

inter-arrival interval. The second "M" signifies exponential service time S. Thus

$$P(S<t)=1-e^{-\mu t}, \quad (t\geq 0) \tag{6.3.2}$$

with similar interpretations for μ. The third symbol in the hieroglyph is an integer which denotes the number of active servers. The hieroglyphs are sometimes expanded to signify that the system is endowed with additional features which the analysis must take account of, e.g. limitation on queue length. In this account no such limitation exists.

Let N(t) be the **system state** at time t after the operation starts. It is an integer-valued random variable indexed on time t. For simplicity we suppose that N(0)=0 and thus we write

$$p_n(t)=P\{N(t)=n\,|\,N(0)=0\}. \tag{6.3.3}$$

From the definition it follows that

$$p_0(0)=1. \tag{6.3.4}$$

Note that N(t) **is independent of queue discipline.**

The first task is to formulate differential-difference equations satisfied by $p_n(t)$. These are usually the Chapman-Kolmogorov equations appropriate to the model. Equations (6.3.5) below are the forward equations. The backward equations are useful in some applications.

Because the driving mechanisms are exponential with parameters λ, μ, it follows that, independently of past history (an important feature of the exponential mechanism), the probability of arrival of a customer in the vanishingly small time interval $(t,t+h)$ is $\lambda h+o(h)$, and the probability of completion in $(t,t+h)$ of a service in progress at time t is $\mu h+o(h)$. $o(h)$ is shorthand for a collection of terms which, when divided by h, tend collectively to zero as h->0. It is shown in the Notes that the set $\{p_n(t):n=0,1,\ldots\}$ of state probabilities satisfies the differential- difference equations

$$\begin{aligned}\dot{p}_0(t)+\lambda p_0(t)&=\mu p_1(t)\\ \dot{p}_n(t)+(\lambda+\mu)p_n(t)&=\mu p_{n+1}(t)+\lambda p_{n-1}(t), \quad (n=1,2,\ldots).\end{aligned} \tag{6.3.5}$$

Dots, as usual, denote differentiation with respect to time.

It can be shown, by a variety of methods, often using the Laplace transformation, but also by direct probability argument that

$$p_n(t)=\frac{e^{-\nu t}}{\mu t}r^n\sum_{k=n+1}^{\infty} r^{-k/2}kI_k(\omega t)$$

with

$$\nu=\lambda+\mu, \quad \omega=2\sqrt{(\lambda\mu)}, \quad r=\frac{\lambda}{\mu}. \tag{6.3.6}$$

Here

$$I_n(z)=\sum_{m=0}^{\infty}\frac{(\frac{z}{2})^{n+2m}}{m!(m+n)!}$$

is the modified Bessel function of the first kind, order n and argument z. Recently attention has been paid to more readily computable forms one of which, due to Sharma [2], but foreshadowed by a number of others, is

$$p_n(t)=(1-r)r^n+e^{-vt}r^n\sum_{m=0}^{\infty}\frac{(\lambda t)^m}{m!}\sum_{k=0}^{m+n}(m-k)\frac{(\mu t)^{k-1}}{k!} \tag{6.3.7}$$

This can be obtained by arguing that if the state is n at time t, then, during (O,t) there may have been $m+n$ arrivals and m service completions ($m=0,1,\ldots$). An alternative form also found convenient for computation is given in [3]. Some values of $p_0(t)$ for various values of the **congestion index** $r=\lambda/\mu$ are given in Table 6.3.1. It is seen that as long as $r<1$, p_0(t) appears to tend, as $t->\infty$, to a nonzero limit, but that, when $r\geq 1$, it appears to decrease towards zero. Is this to be expected?

The answer is affirmative, for if p_n(t) tends to a limit π_n independent of t, the equations satisfied by the π_n must be those satisfied by p_n(t) with zero derivatives. It can thus be verified that $\pi_{n+1}=r\pi_n=r^n\pi_0$, and since by total probability the sum of the π_n over $n\geq 0$ is unity it follows that

$$\frac{1}{\pi_0}=1+r+r^2+\ldots$$

The series on the right is infinite when $r\geq 1$, implying in that case that $\pi_0=0$: otherwise $\pi_0=1\text{-}r$, consistent with the numerical impression given by the table.

The régime that comes about ultimately when $r<1$ is called the **steady state**, corresponding to **statistical equilibrium.** The steady state probabilities are easily seen to form the geometric distribution

$$\pi_n=(1-r)r^n, \quad (n\geq 0,\ r<1). \tag{6.3.8}$$

Steady state probabilities are used to assess the performance of a system but it must not be forgotten that it is often the run up to a steady state and, indeed, of its antithesis, the congested state, that is important. The British National Health Service is a conglomeration of subsidiary queueing systems which could be aggregated into a single, global system. It is criticised for being in a chronically congested condition, but this can only be the case because, roughly speaking, λ is too big compared with μ, or because μ is too small to cope with λ. Money spent on armies of administrators and management consultants might be more effectively deployed in tackling λ and μ head on.

Table 6.3.1 Values of $p_0(t)$

r	0.1	0.5	0.9	1	2	5
t	$p_0(t)$	$p_0(t)$	$p_0(t)$	$p_0(t)$	$p_0(t)$	$p_0(t)$
0	1	1	1	1	1	1
0.5	0.9614	0.8211	0.7009	0.6737	0.4525	0.1351
1	.9386	.7263	.5594	.5238	.2676	.0320
1.5	.9248	.6701	.4792	.4398	.1803	.0095
2	.9162	.6338	.4272	.3858	.1303	.0031
2.5	.9108	.6086	.3904	.3475	.0983	.0011
3	.9073	.5901	.3627	.3187	.0764	.0004
3.5	.9050	.5761	.3409	.2960	.0606	.00016
4	.9035	.5652	.3231	.2776	.0489	.00006
4.5	.9024	.5564	.3084	.2622	.0399	.00002
5	.9017	.5492	.2958	.2491	.0329	.00001
5.5	.9012	.5432	.2850	.2378	.0274	0
6	.9009	.5382	.2755	.2279	.0229	0
6.5	.9006	.5339	.2671	.2191	.0193	0
7	.9005	.5303	.2596	.2113	.0164	0
7.5	.9003	.5272	.2529	.2043	.0139	0
8	.9001	.5244	.2468	.1979	.0119	0
8.5	.900182	.5220	.2413	.1921	.0102	0
9	.900135	.5199	.2362	.1867	.0088	0
9.5	.900100	.5181	.2316	.1818	.0076	0
10	0.900074	0.5165	0.2273	0.1773	0.0066	0

Useful steady-state facts include the following. The notation is that when $r<1$ and $t\to\infty$, $N(t)\to N$.

The moment generating function $M(\theta)$ is given by

$$M(\theta)=E(e^{N\theta})=\frac{1-r}{1-re^{\theta}}, \qquad (6.3.9)$$

from which, by differentiation and subsequently putting $\theta=0$, we get

$$E(N)=\frac{r}{1-r}, \qquad Var(N)=\frac{r}{(1-r)^2}. \qquad (6.3.10)$$

It is possibly counterintuitive that congestion occurs when $r=1$, that is when the mean rate of demand exactly balances the mean rate of service.

On the other hand, how do E(N) and Var(N) behave when $r>1$? In this case the deterministic counterpart to the stochastic model is helpful. If we denote by x the **deterministic system state, or deterministic mean,** at time t, and by λ, μ the arrival and service-completion rates, then x satisfies the differential equation $dx/dt=\lambda-\mu$, giving

$$x = x_0 + (\lambda-\mu)t \qquad (6.3.11)$$

where x_0 is the initial state. Thus, when $r>1$, x increases like t. It can be seen that this is a guide to what happens stochastically. Denote the stochastic mean by $m(t)$. From the set (6.3.5) it follows that

$$\frac{dm}{dt}=\lambda-\mu+\mu p_0(t), \qquad (6.3.12)$$

which differs from (6.3.11) by just the $p_0(t)$ term, and this diminishes to zero, as will its integral, when $r>1$ and $t\to\infty$. The deterministic model also gives guidance about the long-term behaviour of the variance. For, let $m_2(t)=\Sigma n^2 p_n(t)$: then we can find the following differential equation for m_2;

$$\frac{dm_2}{dt}=\lambda+\mu+2(\lambda-\mu)m-\mu p_0, \qquad (6.2.13)$$

from which it follows that as $t\to\infty$ and with $r>1$,

$$Var\{N(t)\}=m_2(t)-m^2(t)\sim(\lambda+\mu)t, \qquad (6.3.14)$$

suggesting that the standard deviation of $N(t)$ increases like $\sqrt{t}$ in the long run. Thus, with 1 customer arriving on average every five minutes, while it takes twice as long to serve a customer on average, the prospect after an hour is a waiting line of 6 with a standard deviation of $\sqrt{(18)}$.

The corresponding analysis for the two-server system M/M/2 is similar. When the system is empty ($N(t)=0$), and the rule is for a new arrival to choose between the two available servers without preference; if one server is busy he goes to the vacant place and, otherwise, waits for the next vacant server. Corresponding to the set (6.3.5), and with similar notation, we have

$$\dot{p}_0(t)+\lambda p_0(t)=\mu p_1(t)$$
$$\dot{p}_1(t)+(\lambda+\mu)p_1(t)=2\mu p_2(t)+\lambda p_0(t) \qquad (6.3.15)$$
$$\dot{p}_n(t)=(\lambda+2\mu)p_n(t)=2\mu p_{n+1}(t)+\lambda p_{n-1}(t), \quad (n\geq 2).$$

because when two or more customers are present the mean rate of service completion is 2μ. Assuming the existence of a non-zero set of state probabilities π_n as $t\text{-}>\infty$ ($n=0,1,\ldots$) leads to $\pi_1=\lambda\pi_0/\mu, \pi_n=r\pi_{n-1}$ for $n\geq 2$ and $r=\lambda/(2\mu)$ this time, and thus

$$\pi_1=2r\pi_0,\ \pi_2=2r^2\pi_0,\ldots,\pi_n=2r^n\pi_0,\ldots, so\ that$$
$$1=\pi_0\{1+2r(1+r+r^2+\ldots)\},$$
$$giving\ \pi_0(1+\frac{2r}{1-r})=1\ as\ long\ as\ r<1.$$

This leads to

$$\pi_0=\frac{1-r}{1+r},\quad \pi_n=2r^n\pi_0,\ if\ r<1$$

and otherwise 0. (6.3.16)

The moment generating function $M(\theta)$, $E(N)$ and the variance $Var(N)$ are as follows:

$$M(\theta)=\pi_0\frac{1+re^\theta}{1-re^\theta},$$
$$E(N)=\frac{2r}{1-r^2},$$
$$Var(N)=\frac{2r(1+r^2)}{(1-r^2)^2}. \qquad (6.3.17)$$

It is opportune now to compare the performance of the single and double server systems in a simple and preliminary way. Of course we expect a two-server system to produce a better (in some sense) performance than a single-server system when the demand and performance of each server is the same for both systems. But it is also of interest to realise that the two-server system actually performs better when subjected to a load that **gives the same p_0 for both systems.** In other words, it seems to operate more efficiently. We again develop this theme under equilibrium conditions, using $E(N)$ as the yardstick of comparison.

For M/M/1 we have $r=\lambda/\mu$ and $p_0=1\text{-}r$, so that $r=1\text{-}p_0$. Then $E(N)=r/(1\text{-}r)=1/p_0 - 1$. For M/M/2 we have $r=\lambda/2\mu$, $p_0=(1\text{-}r)/(1+r)$ and

$r=(1-p_0)/(1+p_0)$. Thus $E(N)=2r/(1-r^2)=(1/p_0 - p_0)/2$.
The corresponding values of E(N) are compared in Table 6.3.2.
A conclusion is that the two-server system, at each given level of traffic for either system, does a better job at keeping the mean system state down and, in that sense, is a more **efficient** system. Of course, the difference between the systems is very slight when traffic is extremely light and one would not in that case go to the expense of a second service anyway. Table 6.3.3 makes a similar comparison, with λ/μ as independent variable.

The comments on traffic intensity apply only to M/M/1, but the comparison this time emphasises the superiority of M/M/2 for the same fixed demand and service performance for each system.

If servers were added so that no demand had ever to wait we should have a so-called M/M/∞ system for which the steady state values are $p_0=e^{-r}$, $E(N)=r$, with $r=\lambda/\mu$, for any $r>0$. Then $E(N)=-\ln p_0$ and, putting corresponding numerical values alongside those in Table 6.3.2 shows that there is no great gain in internal system efficiency for $p_0>0.3$, but great gains can be made in limiting queue size for large r. The M/M/∞ model is useful to simulate a very large car park, the number of shoppers in a supermarket, even the number of ships in the Straits of Dover. Additional discussion will be given later in the section on interactive models.

Table 6.3.2 Mean system size as function of prob. of empty system

Traffic	p_0	$E(N)$ {M/M/1}	$E(N)$ {M/M/2}	$E(N)$ {M/M/∞}
Heavy	0.1	9	4.95	2.30
..	0.3	2.33	1.52	1.20
Moderate	0.5	1	0.75	0.69
..	0.7	0.43	0.36	0.36
Light	0.9	0.11	0.106	0.105
..	0.95	0.053	0.051	0.051

Table 6.3.3 Mean system size as function of λ/μ

Traffic	λ/μ	E(N) {M/M/1}	E(N) {M/M/2}	E(N) {M/M/∞}
Light	0.1	0.11	0.10	0.1
..	0.3	0.429	0.307	0.3
Moderate	0.5	1	0.533	0.5
..	0.9	9	1.129	0.9
Heavy	0.95	19	1.227	0.95
	1	∞	1.333	1
	1.5	-	3.429	1.5
	1.9	-	35.6	1.9
	2	-	∞	2

6.3.2.2. System Time

We turn to the topic which worries a customer the most: How long am I going to have to wait for my hip replacement, to see the dentist, the consultant, the shopkeeper, to pay the cashier at the supermarket outlet, and so forth? The order in which service is processed has now to be specified. In this discussion it will be first-come, first-served.

It is comparatively easy to derive the steady-state probability distribution of system time in the case of exponentially driven queueing systems. Denote a new arrival's system time by W. This is a real-valued random variable whose density and distribution functions will be written $f(w)$, $F(w)$, $(0<w<\infty)$. For M/M/1, the system time of a new arrival when the discipline is first-come, first-served is composed of the remaining service time of the customer being served plus the sum of the service times of the customers waiting in the queue ahead, plus the new arrival's own service time. Residual service time for an exponential distribution has the same distribution as a complete service time so, if the new arrival finds n customers already present, his own system time will be effectively the sum of $n+1$ service times and this holds for $n\geq 0$. The sum of n exponentials with parameter λ has density

$$\mu\frac{(\mu t)^{n-1}}{(n-1)!}e^{-\mu t} \quad (n\geq 1, t\geq 0)$$

and it follows that

$$f(w)=\mu e^{-\mu w}\sum_{n\geq 0} p_n \frac{(\mu w)^n}{n!}$$

$$=\mu(1-r)e^{-\mu(1-r)w} \quad (w\geq 0,\ r=\frac{\lambda}{\mu}). \qquad (6.3.18)$$

Notice that the distribution is pure exponential with parameter $\mu(1-r)=\mu-\lambda$. Thus E(W)=standard deviation of W and is equal to $1/(\mu-\lambda)$.

The principle is the same for M/M/2 but the calculations a little more complicated. They give

$$f(w)=(p_0+p_1)\mu e^{-\mu w}+\frac{2\mu r p_1}{(1-2r)}\{e^{-\mu w}-e^{-2\mu(1-r)w}\}. \qquad (6.3.19)$$

The first term is the contribution when the new arrival finds a server free, and the second when he does not. The second term is, indeed, the weighted density of the sum of two random variables, one being the actual service time of the new customer and the other being the time to clear the system ahead when the service is full on arrival. It may be helpful to recall that when two customers are in service simultaneously, the time until the first service becomes free has density

$$e^{-\mu t}\mu e^{-\mu t}+e^{-\mu t}\mu e^{-\mu t}=2\mu e^{-2\mu t},$$

the two components being the probability that one service finishes in time t and the other takes longer, taken twice because there are two servers. The mean and variance of the system time for M/M/2 are

$$E(W)=\frac{1}{\mu(1-r^2)},\quad Var(W)=\frac{1-r^2+r^3}{\mu^2(1-r^2)^2},\quad (r=\frac{\lambda}{\mu}). \qquad (6.3.20)$$

I was at the local branch of my bank recently where they had decided to replace the system of two tellers, each with its own queue, with the now familiar system where the two tellers are fed by a single queue with, of course, first-come, first-served discipline. My neighbours in the queue, one in front and one behind, evidently knew each other and spent their queueing time complaining about the inefficiency of the new system. They brought me into the discussion by asking if I did not agree. "Yes," I said, "I do not agree. The Bank has consulted experts who predict that customers will save up to half the time. The fame of the Bank will spread and many new customers will be tempted to transfer their accounts." "Nonsense," said one neighbour: "Onzin" said the other, who turned out to be Dutch. The basis of my confident attitude was the following. If the arrival stream intensity λ is divided equally between the two tellers there are, in effect, two single-server systems with arrival intensity $\lambda/2$ and the mean system time, assuming steady-state conditions, is $1/\{\lambda(1-r)\}$ with $r=\lambda/(2\mu)$. This also assumes that customers do not dart back and forth between queues, which might improve their prospects, but I did not mention that since I felt it would be cheating. With a single queue we have seen that the mean system time is $1/\{\mu(1-r^2)\}$, and the ratio of this to the former is $1/(1+r)$ which approaches 0.5 as r approaches unity.

"However," I asked my neighbours,"what do you think of the idea of making one teller redundant and training the other to work twice as fast?" "Not much," they said simultaneously. "You would be better off," I ventured, "though the Bank would probably be too mean to pay a sufficiently attractive salary." In fact the average system time would be $1/\{2\mu(1-r)\}$, a twofold improvement on two queues serviced by lesser mortals, but still not as good as the gain offered by the system my neighbours were previously rejecting.

6.3.2.3 The Busy Period - M/M/1

The conversation reported above evidently floated through the security grille to a young counter clerk who, when it was my turn to be served, asked me whether my studies had considered the strain on the teller who is required to be alert for long periods, watching for banking errors, and apprehensive about bank raids. I thought quickly what I should tell her.

To recapitulate, a busy period B is of primary concern to the service. It begins when the service, previously idle, first receives a customer and then continues until no more customers are left to be served. So the duration may be the time required to serve one customer, or many. Let M(B) be the number of customers served during a busy period of duration B. Then M is a positive integer-valued random variable, and B is a positive time. We would like to find the joint probability and probability-density function $k_n(t)$ associated with values $t<B<t+dt$, $M(B)=n$.

First, it is clear that such a busy period consists of n consecutive services, one following the other without a break. The probability density that the sum of n service times is t, each exponentially distributed with parameter μ, is $\mu\exp(-\mu t)(\mu t)^{n-1}/(n-1)!$, and one would expect that to figure somehow in $k_n(t)$. Also, a busy period serving n customers requires the arrival of $n-1$ during $(0,t)$, and the probability of this is $\exp(-\lambda t)(\lambda t)^{n-1}/(n-1)!$, **but not necessarily in such a way as to ensure no break.** Nevertheless we expect this term too to enter $k_n(t)$. It turns out that of all the ways in which $n-1$ arrivals can occur in $(0,t)$, just $1/n$ of them favours continuation of the busy period and thus

$$k_n(t) = \mu e^{-(\lambda+\mu)t}\frac{(\lambda\mu t^2)^{n-1}}{n!(n-1)!}. \tag{6.3.21}$$

This is made up as follows:

$$\mu e^{-\mu t}\frac{(\mu t)^{n-1}}{(n-1)!} \rightarrow \textit{the probability}$$

density of n successive services ending at time t;

$$e^{-\lambda t}\frac{(\lambda t)^{n-1}}{(n-1)!} \rightarrow \textit{theprobability}$$

of exactly n arrivals in $(0,t)$;

$$\frac{1}{n} \rightarrow \textit{reduction to ensure the}$$

correct pattern of arrivals.

This argument may be made more plausible by using a "ballot theorem" argument. Feller [4], Chapter III, contains a clear account. Here are the means and variances of B and M.

$$r=\frac{\lambda}{\mu},\ (<1),$$

$$E(B)=\frac{1}{\mu(1-r)},\quad Var(B)=\frac{(1+r)}{[\mu^2(1-r)^3]},$$

$$E(M)=\frac{1}{1-r},\quad Var(M)=\frac{r(1+r)}{(1-r)^3}. \qquad (6.3.22)$$

"Here is something," I thought, "to tell the counter clerk. As long as you are not overwhelmed," I said, "meaning while you are in a steady state, you are as well off as the customers. On average you do not have to serve continuously for longer than a customer spends in the system." "But," said she, peering across the counter at the formulae I had jotted down, "it says there that my variance has the cube of $1-r$ in the denominator. If the statistics course I have been attending means anything I could easily be much worse off than the cusromers, whose variance has only the square of $1-r$ in the denominator." How wrong I was in thinking that she would not be technically minded. "However," I said, "you are paid to give service while the customer essentially pays to wait. Now who is worse off?" She looked doubtful, so I promised that if she would read the article I was writing she would find in the next section a discussion of some mechanisms for tightening up operational efficiency and making life better all round for customers and servers. I told her that I called these **interactive systems**. "Interesting," she said, "as long as the service is not expected to be *hyperactive*."

6.3.3 Some Interactive Single-Server Systems

6.3.3.1 The "correlated" queue

Figure 6.3.1 below illustrates the link between the search problem described in Chapter 5.3 and a queueing system in which service is determined by demand. The server observes interarrival intervals and spends a proportionate time providing the customer's need.

Figure 6.3.1 Illustrating the System-Time Process

(a) $W_n > T_{n+1}$

(b) $W_n < T_{n+1}$

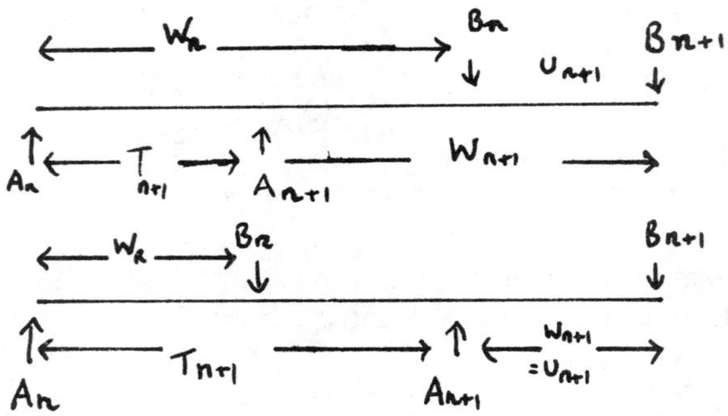

Suppose that successive customers C_n and C_{n+1} arrive at instants A_n, A_{n+1} and complete service at instants B_n, B_{n+1}, respectively. Then A_nB_n and $A_{n+1}B_{n+1}$ denote the system times W_n, W_{n+1} of C_n and C_{n+1}. The interarrival time T_{n+1} "created" by C_{n+1} is A_nA_{n+1}, and the service time U_{n+1} of C_{n+1} is either B_nB_{n+1} or $A_{n+1}B_{n+1}$ according to (a) and (b), and is allocated to C_{n+1} after arrival, so that the server has been able to observe T_{n+1} before starting to serve C_{n+1}. So if, as in this case, the server wishes to allocate time to C_{n+1} in some way dependent on T_{n+1} he could do so. W_{n+1} is given by

$$W_{n+1} = W_n - T_{n+1} + U_{n+1}, \quad \text{if } W_n > T_{n+1},$$
$$= U_{n+1}, \quad \text{if } W_n < T_{n+1},$$

which can also be written

$$W_{n+1} = \max(W_n - T_{n+1} + U_{n+1}, U_{n+1}). \qquad (6.3.23)$$

This formulation was announced by Lindley in 1952 (but see also [1]) and provided the foundation for a range of practically and theoretically interesting studies, including the behaviour of the system time statistics as traffic intensity hovers near congestion level.

Most earlier studies assumed independence of T_n and U_n, but this does not have to be so.In that case we put $U_n = rT_n$ $(0 < r < 1)$ for all n: then (6.3.23) becomes

$$W_{n+1} = \max(W_n - T_{n+1} + rT_{n+1}, rT_{n+1}) \qquad (6.3.24)$$

which may be compared with the formula

$$y_{n+1}=\max(y_n-u\tau_{n+1}+v\tau_{n+1},v\tau_{n+1})$$

developed in Chapter 5.3 in an entirely different context. From the probability density function given there for y_n we can deduce that of W_n. We call this derived queueing system "the correlated queue". When traffic intensity $r<1$ a steady state exists and the probability density function *f(t)* of system time is

$$f(t)=\frac{\lambda}{r}\sum_{n\geq 0} g_n\, e^{-\frac{\lambda s_n t}{r}}$$

with

$$g_n=\frac{(-)^n r^{n(n-1)/2}}{(1-r)^2(1-r^2)(1-r^3)...(1-r^n)} \qquad (n\geq 1, s_n=\sum_{i=0}^{n}\frac{1}{r^i}),$$

and

$$g_0=\frac{1}{1-r}. \qquad (6.3.25)$$

The density is a linear combination of exponentials giving what is often called a hyperexponential distribution. The mean and variance are:

$$E(W)=\frac{r}{\lambda}\sum_{n\geq 0}\frac{1}{s_n}, \quad Var(W)=(\frac{r}{\lambda})^2\sum_{n\geq 0}\frac{1}{s_n^2}. \qquad (6.3.26)$$

The following Table 6.3.4 compares steady state E(W) and Var(W) for M/M/1 with $E_c(W)$, $Var_c(W)$ for the correlated system under a range of traffic intensities *r*. Note that for the latter the probability p_0 of an empty system *just before an arrival* is 1-*r*, as for M/M/1 *at any time*. The time scale is fixed by setting $\lambda=1$, corresponding to unit mean interarrival interval. In this case, for M/M/1,

$$E(W)=\frac{r}{1-r}, \quad Var(W)=(\frac{r}{1-r})^2.$$

Table 6.3.4 Comparison of means and variances of correlated with M/M/1 system

	M/M/1		Correlated	
r	E(W)	Var(W)	E_C(W)	Var_C(W)
0.1	0.1111	0.0123	0.1101	0.0101
.3	0.4286	0.1837	.3968	.0952
.5	1	1	0.8033	.2843
.7	2.333	5.444	1.427	0.6132
.9	9	81	2.709	1.164
.95	19	381	3.470	1.365
0.99	99	9801	5.153	1.572

The improvement conferred by correlation is extremely gratifying under very heavy traffic and the reduction in variability quite startling. "But Hallo!" said the statistically inclined lady from the bank; "Could you really operate such a system?" "Maybe not exactly," I replied, "but it shows what cooperation is capable of and provides a challenge to find a way of implementing the spirit, if not the letter." "Unfortunately," I added, "no-one has so far found a way of dealing with the analysis of a general busy period. We understand the initial one but have to rely on simulation methods for the rest. The results do look promising and I suspect there is something in it for the service as well."

6.3.3.2 Some state-dependent systems

Here we deal, not with government agencies, but with exponential systems where the mean arrival and service rates are subscripted by the system size n,(thus λ_n, μ_n, to indicate that, independently of time, the probability of an arrival in $(t, t+h)$ when the system state $N(t)=n$, is $\lambda_n h+o(h)$ $(n\geq 0)$, and the probability of a service completion is $\mu_n h+o(h)$, $(n\geq 1)$. λ_0 can be defined as zero if needed. Notice that the "Markov lack of memory property" holds. The forward Chapman-Kolmogorov equations are

$$\dot{p}_0(t)+\lambda_o p_0(t)=\mu_1 p_1(t),$$
$$\dot{p}_n(t)+(\lambda_n+\mu_n)p_n(t)=\mu_{n+1}p_{n+1}(t)+\lambda_{n-1}p_{n-1}(t), \quad (n\geq 1). \quad (6.3.27)$$

In some cases these can be solved to give the time-dependent forms of the system-state probabilities $p_n(t)$.

The mechanisms are useful for modelling:
(a) modulation of server responsiveness to demand based on observation of the system state N(t). If μ_n increases with n the service is working faster (cooperation), while if it decreases the service slows down (a possible indication of fatigue, neurosis, obstructiveness...);
(b) modulation of demand in response to observation of N(t). Decrease of λ_n with n indicates reluctance to join a crowd and conversely, when it increases with n, potential customers are encouraged to join, as if there is some highly desirable commodity on offer. We shall discuss the following briefly:

Model CS: Cooperative service with $\mu_n = \mu n$ $(n \geq 1)$, $\lambda_n = \lambda$ (constant) and, of course, λ, μ positive.

Model RD: Reluctant demand, with $\lambda_n = \lambda/(n+1)$, $\mu_n = \mu$ (constant).

Model C: The simultaneous combination of $\lambda_n = \lambda/(n+1)$, $\mu_n = \mu n$.

For reference to extensive theoretical work see [8].

Time-dependent probabilities and their steady-state counterparts are as follows:

Model CS: The Cooperative Server
The mechanism is equivalent to M/M/∞, mentioned in 6.3.2. For an initially empty system (N(0)=0, $p_0(0)=1$)

$$p_n(t) = \frac{[r(1-e^{-\mu t})]^n}{n!} e^{-r(1-e^{-\mu t})}, \quad (r = \frac{\lambda}{\mu}), \qquad (6.3.28)$$

which means that N(t) has a Poisson distribution. The steady-state probabilities are

$$p_n = e^{-r}\frac{r^n}{n!}. \qquad (6.3.29)$$

Model RD: Reluctant Demand
Sharma (private communication) gives the formula

$$p_n(t) = \pi_n + \sum_{m=1}^{\infty} A_{nm} e^{-\alpha_m t}, \quad \pi_n = e^{-r}\frac{r^n}{n!}. \qquad (6.3.30)$$

This isolates the steady-state component which is the same as (6.3.26) above for the CS model. The constants in the sum are complicated. The result was obtained by analysis of a finite-sized system, letting finally the size tend to infinity. The important fact is that an accessible computational formula is available for $p_n(t)$, that it isolates the steady state component, and that it exposes the remaining form as hyperexponential.

Model C: Combination of CS and RD

The following formula is given in [9]. With an initially empty system, $p_n(t)$ is the coefficient of x^n in

$$G(x,t)=\frac{I_0[2\sqrt{rx}]}{I_0[2\sqrt{\{r(1+(x-1)e^{-\mu t})\}}]},$$

and with $r=\lambda/\mu$

$$p_0(t)=\frac{1}{I_0[2\sqrt{r(1-e^{-\mu t}}]}. \qquad (6.3.31)$$

One measure of performance of these systems compared with M/M/1 is $p_0(t)$. Table 6.3.1 gives values for M/M/1 and a range of $r=\lambda/\mu$ and t. Calculations of the same quantity from (6.3.25), (6.3.27) and (6.3.28) show that:

(i) all are superior to M/M/1 in the sense that $p_0(t)$ is uniformly larger;

(ii) Model RD is better for finite t than CS, and C is better than RD, both **for all t.**

(ii) can also be tested by comparing the limiting values of $p_0(t)$ as $t\text{->}\infty$. Denoting this limit by p_0, we note that both CS and RD give $p_0=\exp(-r)$ which, as $r\text{->}0$, is approximately $1-r+r^2/2$, greater than the M/M/1 value $1-r$. Also, the first terms of $I_0(2\sqrt{r})$ are $1+r+r^2/4$, so that, for C, p_0 is approximately $1-r+3r^2/4$, exhibiting superiority over RD. On the other hand, as $r\text{->}\infty$,

$$I_n(z)\sim\frac{e^z}{\sqrt{2\pi z}}$$

which means that as $r\text{->}\infty$, for C,

$$p_0\sim 2\pi^{1/2}r^{1/4}e^{-2r^{1/2}},$$

the ratio of which to e^{-r} is

$$2\pi^{1/2}r^{1/4}e^{-1}e^{(r^{1/2}-1)^2},$$

which is clearly >1. On the basis of this crude comparison we conclude that C is better than RD is better than CS is better than M/M/1 in keeping the system state within bounds. For more detailed numerical comparisons see [8].

The following Table, extracted from that Reference, gives the theoretical values of mean system time W', mean busy period length B', and p_0, to enable the reader to make more detailed comparisons.

Model	p_0	**W'**	**B'**
M/M/1	$1-r$	$[\mu(1-r)^{-1}$	W'
CS	e^{-r}	μ^{-1}	$(e^{r}-1)/\lambda$
RD	e^{-r}	$r[\mu(1-e^{-r})]^{-1}$	$(e^{r}-1)/\lambda$
C	$[1-I_0(2\sqrt{r})]^{-1}$	μ^{-1}	$[I_0(2\sqrt{r})-1]/\lambda$

6.3.4. The Ugly Side of Priority

You have dropped a sledgehammer on your foot and are waiting to be seen in the Casualty Department of your local hospital since you are in pain, find it hard to walk, and fear that you may have broken a toe or two. You have been waiting for an hour and it looks as though your turn will be next, but then a patient is brought in with blood streaming from a cut cheek and you lose your turn. After another ten minutes you are told that you will be next so you rehearse the speech you intend to make to the doctor, but again your prospects are demolished by the arrival of a stretcher case bearing an individual evidently in a **very bad way**, who is rushed in instead of you.

Experiences of this kind are common enough. The customers with priority have relatively little time to wait, but this is at the expense of those with lower priority.

In what follows we shall, without discussing justification, turn anecdote into numbers in the context of M/M/1 and what is called **non-pre-emptive priority**. This means that a customer with priority has to wait until the service is next empty, whereupon he,as well as all other priority customers who arrive during the service of their predecessors, totally absorb the service facility until there are no more priority customers left. Thus, non-priority customers have no further turn until the busy period initiated by the arrival of the priority customer and his successors is terminated. The fact that busy periods form a component of the system time of non-priority customers is bad news for them since busy periods, as we saw, have unfavourable dispersion characteristics as traffic intensity approaches saturation.

To quantify these statements we assume that customers are divided into two classes:priority (P), and non-priority (NP). P customers constitute a proportion α of the exponential arrival stream (mean rate parameter λ) and $0<\alpha<1$. NP customers are a proportion $\underline{\alpha}$ $(=1\text{-}\alpha)$. We also assume that service time for all customers is exponential with mean rate parameter μ. The priority is non-preemptive, meaning that service continues in sequence on NP customers as long as no P is present. When a P arrives, if a NP customer is being served, the P waits until it is finished and then takes over the service for himself and any P successors until there are no more waiting. It is convenient to think of the waiting customers as divided into a P-queue and a NP- queue. A very brief sketch of the theory is given in the Notes, and [6] may be consulted for more detail.

We now look at some means and standard deviations of system time for P and NP customers under a variety of conditions, one of which is that the system is in equilibrium. The values are taken from tables in [6]. It is assumed that FCFS discipline applies to both P and NP queues. The time scale is fixed by setting $\mu=1$.

Table 6.3.5 Mean equilibrium system times of P- and NP-streams against $r=\lambda/\mu$ and P-ratio α

	P	NP	P	NP	P	NP	P	NP
α	$r=$	0.1	$r=$	0.5	$r=$	0.9	$r=$	0.95
0.1	1.101	1.112	1.526	2.053	1.989	10.89	2.05	22.0
0.5	1.103	1.117	1.667	2.333	2.636	17.36	2.81	37.2
0.9	1.110	1.122	1.909	2.818	5.737	48.37	7.55	132
M/M/1	1.111		2		10		20	

Table 6.3.6 Standard deviations of system times of P- and NP-streams

	P	NP	P	NP	P	NP	P	NP
α	$r=$	0.1	$r=$	0.5	$r=$	0.9	$r=$	0.95
0.1	1.093	1.114	1.353	2.107	1.482	11.0 7	1.489	22.2
0.5	1.100	1.128	1.528	2.742	2.067	19.4 2	2.149	39.7
0.9	1.109	1.144	1.865	4.041	5.331	69.8	6.960	172
M/M/1	1.111		2		10		20	

For comparison, the values for non-priority M/M/1 are given at the column feet. The difference is that here the time scale is set by $\mu=1$, whereas in Table 6.3.4 we put $\lambda=1$. A priority régime is obviously more practical, as well as tempting, when traffic is heavy so we can concentrate on performance for $r=0.9$ and 0.95. For example, at $r=0.9$,when there is an equal mixture of P and NP customers ($\alpha=0.5$), a P-customer can expect system time of 2.6 units, on average, with a standard deviation of 2.07, compared with 10 units under straight M/M/1. On the other hand, the NP-customer will spend on average 17.4 units in the system with standard deviation 19.4. This unfavourable comparison worsens for NP's as the proportion of P's increases, and as r increases.

It is also of interest to notice that if, instead of a priority system, the correlated queue mechanism could be implemented, **all customers** could expect to spend an average of 3.01 time units in the system with a standard deviation of 1.2 units.

Does this tell us something about cooperative systems? Of course, in the case of genuine need, priority has to be considered, but perhaps a cooperative system which harmonises mean service rate with demand could reduce the need for priority so often and thereby the unfortunate effect it can have on the non-priority stream. In the case of traffic, the example with which we started, traffic lights can provide a palliative to long queues. An alternative discipline, discussed in [6] in some detail, is alternating priority: here, as soon as one stream is served to exhaustion, the other stream is allowed to take over and to continue until no-one is left, and so on.

Some highlights of the analysis are given in the Notes, but the reader who wishes to pursue the topic is recommended to consult [6].

6.3.5 Notes

6.3.5.1. State Equations for M/M/1 and M/M/2

We formulate an expression for $p_n(t+h)=P\{N(t+h)=n|N(0)=0\}$ in terms of the probabilistically non-negligible states at time t. This means that, because probabilities of state transitions greater than 1 and of multiple events (like simultaneous arrival and end of service) over a small interval $(t,t+h)$ have magnitude o(h), we need consider only the possibilities $N(t)=n\text{-}1,n,n+1$ when $n>0$, and otherwise $N(t)=0$ or 1. The exponential mechanisms imply that arrival and completion of service in progress in $(t,t+h)$ have probabilities $\lambda h+o(h)$ and $\mu h+o(h)$, respectively. For M/M/2, when the system state is 2 or more, the infinitesimal probability of a service completion is $2\mu h+o(h)$ since there are two servers collectively performing twice as fast.

For M/M/1 and $n\geq 1$,

$$p_n(t+h)=p_{n-1}\lambda h+p_n(t)\{1-(\lambda+\mu)h\}+p_{n+1}(t)\mu h+o(h).$$

For M/M/2 and $n \geq 2$,

$$p_n(t+h)=p_{n-1}(t)+p_n(t)\{1-(\lambda+2\mu)h\}+p_{n+1}(t)2\mu h+o(h).$$

For M/M/1 and M/M/2 we get,

$$p_0(t+h)=p_0(t)(1-\lambda h)+p_1(t)\mu h+o(h),$$

and, finally, for M/M/2 and $n=1$,

$$p_1(t+h)=p_0(t)\lambda h+p_1(t)\{1-(\lambda+\mu)h\}+p_2(t)2\mu h+o(h).$$

The standard technique is to expand the left hand side of each equation for small h by Taylor's Theorem up to the o(h) terms, to cancel a term from right and left sides, to divide through by h and, finally to let h->0. This delivers (6.3.5) and (6.3.15).

6.3.5.2. Non-Pre-emptive Priority Theory in the Context of M/M/1 A Sketch

As usual, the arrival stream is exponential with mean rate λ, and this is composed of a proportion α of priority (P) customers who may be thought of as forming a P-queue for service. The remaining proportion $\underline{\alpha}(=1-\alpha)$ consists of non-priority (NP) customers. The priority discipline is as follows. The service attends exclusively to P-customers as long as they are present. Then the service turns to NP-customers and continues as long as no P-customer arrives. When a P-customer does arrive and finds NP-customers present he waits until current service terminates and then occupies the service both for himself and any other P-customers who arrive during his service. Thus, the commencement of service of a P-customer initiates a P-customer busy period. It is also assumed that all customers receive exponential service time with mean rate μ and that the system is in overall equilibrium. $r=\lambda/\mu$ is the traffic parameter. Let X and Y denote, respectively, the numbers of P-, and NP-customers present (X,Y$\geq$0). It is convenient also to introduce a random variable Z which takes the value 1 when a P-customer is being served, and 0 when it is an NP-customer in service. Z is not defined when X=Y=0. The basic probabilities are

$$a_{mn}=\mathrm{P}(\mathrm{X}=m,\mathrm{Y}=n,\mathrm{Z}=1),(m\geq 1,n\geq 0);b_{mn}=\mathrm{P}(\mathrm{X}=m,\mathrm{Y}=n,\mathrm{Z}=0),(m\geq 0,n\geq 1)$$

The system state

$$\mathrm{N}=\mathrm{X}+\mathrm{Y},\quad (\mathrm{N}\geq 0),$$

and

$$\pi_n=\mathrm{P}(\mathrm{N}=n)=(1\text{-}r)r^n,\ (n\geq 0). \qquad \text{(See 6.3.1).}$$

It follows from the definitions that

$$\pi_n = \sum_{m=1}^{n} a_{m,n-m} + \sum_{m=0}^{n-1} b_{m,n-m}.$$

It is worth interpolating here a brief note on the **Method of Balance**, a conservation principle useful in developing the equilibrium state probabilities for M/M/ queues. Take M/M/1 as an introductory example.

Consider a row of boxes labelled $0,1,\ldots,n,\ldots$, corresponding to values of the state N, and imagine each to contain "heaps" of probability - something like silver sand - so that box n contains a heap whose volume is π_n. As time passes probability shifts from each box to adjacent boxes (box in the case $n=0$) so that at any time there is flow of amount $\lambda\pi_n$ from box n into box $n+1$, and a flow of amount $\mu\pi_n$ into box n-1. If the system is in equilibrium the total flow out must be balanced by flow into box n. This is $\mu\pi_{n+1}$ from box $n+1$ and $\lambda\pi_{n-1}$ from box n-1. The diagram illustrates this for $n \geq 1$, and for the special case of $n=0$.

Diagram 1

We have

OUT		IN	
$(\lambda+\mu)\pi_n$	$=$	$\mu\pi_{n+1} + \lambda\pi_{n-1}$	$(n \geq 1)$

and

$$\lambda\pi_0 = \mu\pi_1,$$

which, together lead to the known results $\pi_n=(1-r)r^n$, $(r=\lambda/\mu, n\geq 0)$.

In the case of the priority queue the states are (X,Y) with probabilities a_{mn}, b_{mn} according to who is in service.
First we have

Diagram 2

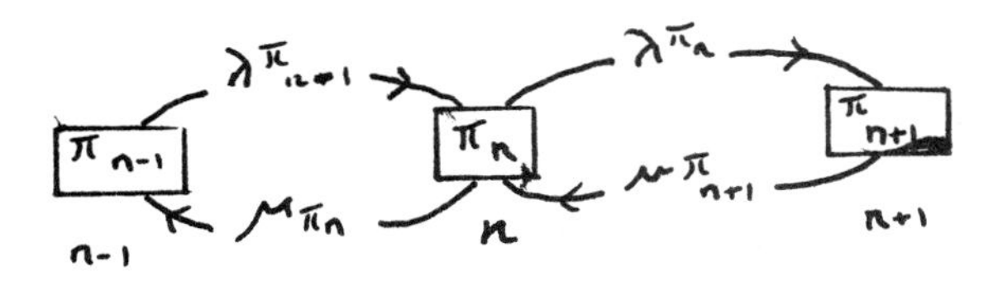

which gives

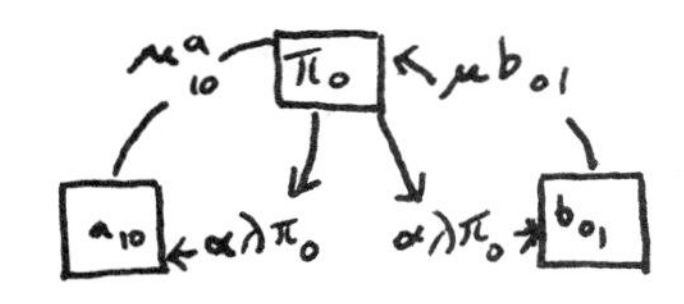

$$(\alpha\lambda+\underline{\alpha}\lambda)\pi_0=\mu a_{10}+\mu b_{01},\ or$$
$$a_{10}+b_{01}=r\pi_0=\pi_1.$$

Passing next to state N=1 we have flows as indicated below:

Diagram 3

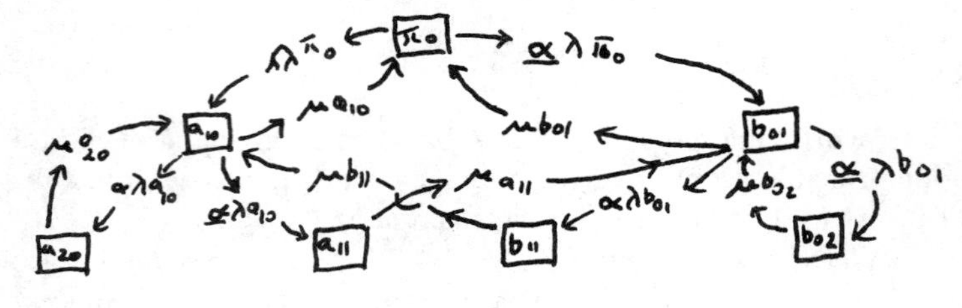

Thus,

$$(\alpha\lambda+\underline{\alpha}\lambda+\mu)a_{10}=\mu a_{20}+\mu b_{11}+\alpha\lambda\pi_0$$
$$(\alpha\lambda+\underline{\alpha}\lambda+\mu)b_{01}=\mu a_{11}+\mu b_{02}+\underline{\alpha}\lambda\pi_0.$$

The following equations can be written down by using the same technique.

$$(1+r)a_{n0}=a_{n+1\ 0}+b_{n1}+\alpha r a_{n-1\ 0}\quad (n\geq 2)$$
$$(1+r)a_{n-1\ 1}=a_{n1}+b_{n-1\ 2}+\alpha r a_{n-2\ 1}+\underline{\alpha} r a_{n-1\ 0}\quad (n\geq 3)$$
$$\cdots\cdots$$
$$(1+r)a_{2\ n-2}=a_{3\ n-2}+b_{2\ n-1}+\alpha r a_{1\ n-2}+\underline{\alpha} r a_{2\ n-3}$$
$$(1+r)a_{1\ n-1}=a_{2\ n-1}+b_{1\ n}+\underline{\alpha} r a_{1\ n-2}$$
$$(1+r)b_{0n}=a_{1n}+b_{0\ n+1}+\underline{\alpha} r b_{0\ n-1}\quad (n\geq 2)$$
$$(1+r)b_{1\ n-1}=\alpha r b_{0\ n-1}+\underline{\alpha} r b_{1\ n-2}\quad (n\geq 3)$$
$$\cdots\cdots$$
$$(1+r)b_{n-2\ 2}=\alpha r b_{n-3\ 2}+\underline{\alpha} r b_{n-2\ 1}$$
$$(1+r)b_{n-1\ 1}=\alpha r b_{n-2\ 1},\quad (n\geq 2).$$

With some labour it is possible to show that the system times W_P, W_{NP} of priority and non-priority customers have the following statistical properties. The probability density function $g_P(t)$ of W_P is

$$g_P(t)=\mu\{\frac{(1-\alpha r)}{\alpha}e^{-(1-\alpha r)t}-\frac{\underline{\alpha}}{\alpha}e^{-\mu t}\}$$

from which

$$E(W_P)=\frac{1+\underline{\alpha}r}{\mu(1-\alpha r)},Var(W_P)=\frac{1+2\underline{\alpha}r-r^2(1-\alpha^2)}{\{\mu(1-\alpha r)^2\}}.$$

W_{NP} is more complex. The probability density function $g_{NP}(t)$ has Laplace transform

$$\gamma_{NP}(z) = \int_0^\infty e^{-zt} g_{NP}(t)dt$$

given by

$$\gamma_{NP}(z) = \frac{\mu\pi_0}{(\mu+z)\{1-r\kappa(z)\}}$$

where $\kappa(z)$ is the Laplace transform of the P-customer busy period duration density (obtainable from Section 6.3.2), namely,

$$\kappa(z) = \frac{Z-R}{2\alpha\lambda} \quad ,Z=z+\mu+\alpha\lambda, \quad R^2=Z^2-4\alpha\lambda\mu.$$

The distribution function $G_{NP}(t)$ can be expressed in terms of Bessel functions, but the moments can be obtained by differentiation of the Laplace transform. They are

$$E(W_{NP}) = \frac{1-\alpha r+\alpha r^2}{\mu(1-r)(1-\alpha r)},$$

$$Var(W_{NP}) = \frac{1}{\mu^2} + \frac{r(2-2\alpha r+\underline{\alpha} r+\alpha\underline{\alpha} r^2)}{\mu^2(1-\alpha r)^4(1-r)} + \frac{\underline{\alpha}^2 r^2}{\mu^2(1-\alpha r)^4}\left(\frac{\alpha}{\underline{\alpha}^2} + \frac{r}{(1-r)^2}\right).$$

One of the depressing features of W_{NP} is that its variance depends on $(1-\alpha r)^{-4}$. More detail on the above will be found in [6].

References

1.Brockmeyer,E., Halström,H.L., & Jensen, A. (1948) *The life and works of A.K. Erlang*. The Copenhagen Telephone Company. (For the information of bibliophiles, this is a rare book.)

2.Sharma, O.P. (1990) *Markovian queues*. Ellis Horwood. Chichester.

3. Conolly, B.W. and Langaris, Christos.(1993) On a new formula for the transient state probabilities for M/M/1 queue... J.Appl. Prob., 30, 237-246.

4. Feller, W. (1957) *An introduction to probability theory and its applications*. Wiley.

5. Conolly, B.W. (1975) *Queueing systems*. Ellis Horwood. Chichester.

6.Conolly, Brian, and Pierce, John G. (1988) *Information mechanics.* Ellis Horwood. Chichester.

7.Conolly, Brian (1990) Queueing and Related Theory. Chapter 12 in *Handbook of applicable mathematics - Supplement.* Wiley.

8.Chan, J. and Conolly, B.W. (1978) Comparative effectiveness of certain queueing systems. Comput. and Ops. Res.,**5**, 187-196.

9. Conolly, Brian (1983) More transient results for generalised, state-dependent Erlangian queues. Adv. Appl. Prob.,**15**,688-690.

6.4 UNCERTAINTY IN STOCK CONTROL

6.4.1. Basic Model

Stock-, or inventory-control at its most basic is easy to formulate, to model, and to understand. The concept is that a controller is in charge of a warehouse containing a range of commodities C_i $(i=1..n)$ which are to be supplied on demand to customers. The demand for C_i is a fixed number D_i of units per unit time (e.g. a week), and it is the controller's job to see that customers get what they want, when they want it, subject to thrifty operation of the system. We now drop subscripts and deal with a particular commodity C for which the demand is D units per unit time. At its simplest the costs of operation are:

(i) **A fixed transport cost £K independent of the number of items carried.** It would therefore seem prudent to order as many as possible at any one time.

(ii) **A holding charge** £h per unit of C per unit time. This is the price the controller must pay for using space in the warehouse and for a realistic inventory-control system the contention for space by different commodities must be looked at too. To keep the holding charge down the controller should keep as few items as possible in stock consistent with satisfying customers, and this is in opposition to the natural inclination to order as many items as possible in order to minimise the transport charge per item.

(iii) **Penalty charge for "Stockout"**. Stockout is the state of being out of stock and therefore unable to satisfy demand immediately. There are various disadvantages, not the least being the potential loss of customers' good will. The system must accordingly be made to pay. In the interest of simplicity it is supposed that customers who can not be supplied immediately are prepared to await the next delivery, but, to punish the operator, a penalty charge of £p per unit quantity of C per unit time is levied. This is a kind of negative holding, or rental, charge and is calculated in the same way.

Operating time is divided into cycles of length T time units. At the beginning of a cycle a quantity DT of items C is delivered, exactly enough to satisfy the assumed average level of demand over the cycle time. Assuming that stockout is allowed, the waiting customers' demands are first met and it follows that the average stock level will fall to zero again before the end of the current cycle. We now calculate the cost of operation on the assumption that stockout of maximum amount $(D\text{-}d)T$ $(0 \leq d \leq D)$ is allowed. The diagram shows the stock level on the vertical axis during a typical cycle which begins at A and finishes at B, time T later. The horizontal axis is time.

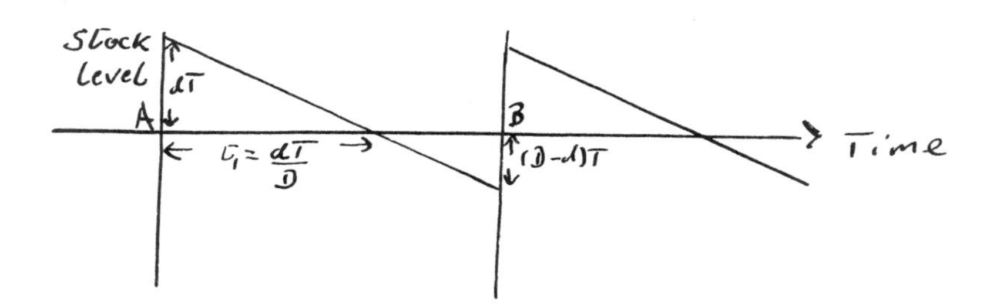

It is seen that stockout occurs at time t_1 such that $Dt_1 = dT$. The holding charge is the average positive stock level multiplied by t_1, multiplied by h,namely $\frac{1}{2}(dT)t_1 h = \frac{1}{2}(dT)^2 h/D$, while the penalty charge incurred for stockout during the period (t_1, T) is calculated in the same way and gives $\frac{1}{2}(D\text{-}d)^2 T^2 p/D$. Thus, taking into account delivery cost K, the cyclic cost R of operating this system, ignoring questions of profit which are irrelevant to R, is

$$R = K + \frac{T^2}{2D}\{d^2 h + (D-d)^2 p\}.$$

The cost of the operation per unit time is then

$$r = \frac{R}{T} = \frac{K}{T} + \frac{1}{2}BT$$

where

$$B = \frac{d^2 h + (D-d)^2 p}{D},$$

and it is the operator's first problem, knowing D, K, p, h to choose replenishment cycle time T, and d, which governs maximum permitted stockout level, so that r is minimized. We see first that r is a convex-down function of T such that, whatever permissible choice is made for d, it has a minimum in $(0,T)$. This occurs for $T = T^*$ satisfying $\mathrm{d}r/\mathrm{d}T = 0$ and thus gives

$$T^* = \sqrt{2K/B}.$$

For this value of T

$$r = \sqrt{2KB}$$

and this will be least when d is chosen to minimise B. It is easily seen that the value d^* is given by

$$d^* = \frac{Dp}{p+h},$$

giving

$$B = \frac{Dhp}{p+h}.$$

Thus, the minimum operational cost r^* per unit time is

$$r^* = \sqrt{\frac{2KDhp}{p+h}}.$$

An interesting feature of this formula is that it permits the conclusion to be drawn that, under the conditions stated, it is **always** more effective to allow stockout. For, to prohibit stockout means making it infinitely expensive. Thus we look at the limit as $p \to \infty$ and observe that

$$r^* \to \sqrt{2KDh},$$

which is greater than the value of r^* when p is finite. The conclusion is,nevertheless, predicated on the willingness of unfulfilled customers to await the next replenishment.

From r^* we can calculate the optimal reorder cycle time T^*. We get

$$T^* = \sqrt{\frac{2K(p+h)}{Dhp}},$$

and the corresponding size of the replenishment order is

$$DT^* = \sqrt{\frac{2KD(p+h)}{hp}}.$$

This is called the **E**conomic **O**rder **Q**uantity (EOQ) and has a vogue in the milieu of management mathematics.

6.4.2. Coping with uncertain Delivery

The model above makes no provision for uncertainty. This can be in demand, delivery, or both. First we examine the consequence of delivery unreliability which may arise through, for example, adverse weather, or labour disputes. Thus, we adopt the new assumption that, although scheduled to arrive regularly at intervals T, replenishment stock may be delayed. In this case, the ideal is to deliver it plus any backlog at the end of the next cycle. But this delivery may also be delayed and the same rule will apply, namely to deliver new stock plus backlog at the next replenishment opportunity. And so on. What difference does this make to the choice of T?

To simplify the analysis we suppose that no planned stockout is allowed for. This means that in principle the stock level is topped up to DT at the beginning of each replenishment cycle. Let a be the probability that the normal order DT plus any backlog is delivered and b that it is not, so that $a+b=1$. Let X_n be the stock level immediately before the n-th replenishment instant, with $X_1=0$,and let X_n' be the level immediately after. Let

$$u_r^{(n)}=P[X_n=-rDT],(r=0,1,...,n-1),$$
$$\text{and}$$
$$v_r^{(n)}=P[X_n'=-rDT],(r=-1,0,...,n-1).$$

Then, since $X_n'=DT$ if and only if a delivery occurs,and since $X_n'=-rDT$ if and only if $X_n=-rDT$ and no delivery occurs, we have

$$v_{-1}^{(n)}=a, \qquad v_r^{(n)}=bu_r^{(n)}.$$

Also, since X_n has to be an amount DT less than X_{n-1}' because of the demand over the previous cycle, we have

$$u_r^{(n)}=v_{r-1}^{(n-1)} \qquad (r=0,1,...,n-1),$$

giving

$$u_r^{(n)}=ab^r, \qquad (r=0,1,...,n-2)$$
$$u_{n-1}^{(n)}=b^{n-1},$$

which may be proved by induction. A check is that

$$S_n=\sum_{r=0}^{n-1} u_r^{(n)}=a+bS_{n-1},$$

so that if $S_{n-1}=1$ then so also is S_n, and we have $S_1=1$.

The probabilities $u_r^{(n)}$ are needed to evaluate the cost of the operation, to which we now proceed. We shall evaluate the **long-run** cost R as if the system had been in operation for a very long time, thus using the limiting values $u_r=\lim u_r^{(n)}$ as $n->\infty$. This gives

$$u_r = ab^r \quad (r\geq 0),$$

a **geometric** distribution.

The **transport** element K of cost is assumed to remain constant, and here we should advise the reader that he should feel free to vary any of the assumptions built into the analysis. The result may be awkward expressions, but these do have to be faced in real life, though they do not make much fun in an essay. In this instance it might be felt realistic to build in an increase in transport costs in proportion to the amount the policy requires the carrier to handle. On the other hand, the fault of non-delivery might be laid at the door of the transport facility and in that case it is reasonable that there should be no penalty to the customer.

The **holding charge** H arises only when $X=0$ (meaning that stock level has decreased from DTT to 0 during the preceding cycle). The associated probability is $u_o=a$. The component of R is

$$H = \tfrac{1}{2}DT^2ha.$$

The **stockout penalty** P is calculated as the average negative stock level X multiplied by Tp. Suppose that $X=-rDT$, with probability ab^r. The stock level has fallen from $-(r-1)DT$ to $-rDT$ over the preceding cycle. The mean deficit is $\tfrac{1}{2}DT(r+r-1)=DT(r-\tfrac{1}{2})$ and the product with T and p gives $DT^2p(r-\tfrac{1}{2})$. Then P is given by

$$P=DT^2p\sum_{r\geq 1}(r-\frac{1}{2})u_r=DT^2pb(\frac{1}{a}-\frac{1}{2}).$$

Thus

$$R = K + H + P$$

and

$$r = K/T + BT$$

where

$$B=Dpb(\frac{1}{a}-\frac{1}{2})+\frac{1}{2}Dha.$$

This leads, as before, to optimal T^* given by

$$T^{*2}=\frac{K}{B}=\frac{2K}{D\{pb(\frac{2}{a}-1)+ha\}},$$

and optimal cost r^* per unit time is

$$r^*=2\sqrt{BK}.$$

The reader will want to check that this expression reduces correctly for certain delivery and may wish further to check what uncertainty does to cost per unit time. We shall turn next to uncertainty in estimation of demand.

6.4.3. Coping with uncertain Demand

To his dismay the controller finds one day that he had underestimated demand and that it is actually $D(1+\xi), \xi>0$. This controller had hitherto operated on a non-stockout basis, so d in the analysis above is zero. The optimal cycle time T^* and cost per unit time r^* are connected by

$$T^*=\sqrt{\frac{2K}{Dh}}=\frac{r^*}{Dh}.$$

However, underestimation creates stockout at time t_1 given by

$$D(1+\xi)t_1=DT.$$

The operating cost is thus

$$R=K+\frac{1}{2}DTt_1h+\frac{1}{2}D\xi Tp(T-t_1)$$
$$=K+\frac{1}{2}\frac{DT^2}{1+\xi}(h+\xi^2p)$$

giving

$$r=\frac{K}{T}+BT, \qquad B=\frac{1}{2}\frac{D(h+\xi^2p)}{1+\xi}.$$

In these formulae T is, of course, T^*.

Some algebra gives the cost r per unit time in terms of the optimal cost r^* before underestimation of demand:

$$r = r^*\{1 + \frac{\xi(\xi p - h)}{2h(1+\xi)}\}.$$

We see that in fact $r<r^*$ if $0<\xi<h/p$ so, in a sense, underestimation has proved advantageous. But this is not the case when demand is overestimated. Suppose that demand is $D(1-\xi)$ with positive ξ. In this case stockout does not arise and a repetition of the now familiar calculations gives

$$r = r^*(1 + \frac{\xi}{2}).$$

The controller observes that overestimation is always costly.

A standard method recommended for probabilistically fluctuating demand is called the (s,S) procedure and the details can be found in reputable textbooks. A sketch is also given in the Notes.

Reference [1] gives an excellent account of the essentials of inventory theory.

6.4.4 Notes: The (s,S) procedure

Let X be the stock level **just before** a replenishment instant, and X' just after. The method proposed to operate the system is to top X up to S (so that $X'=S$) when $X<s$, and otherwise to do nothing. Thus $X'=S$ if $X=S$ or $X<s$ and otherwise $X'=X$. The system and its consequences is driven by a law governing demand. Let, for example, the probability a_r that r items (we are discrete) are demanded during a replenishment cycle be geometric, viz.

$$a_r = (1-a)a^r, \quad (r=0,1,\ldots).$$

The following diagram illustrates the mechanism. A steady state is assumed so that X at A, just before replenishment, if any, has the same distribution (f_n) as X just before the next replenishment at B. Likewise the stock level X' just after replenishment has the same distribution (g_n) at A and B. A and B are separated by the replenishment cycle time T.

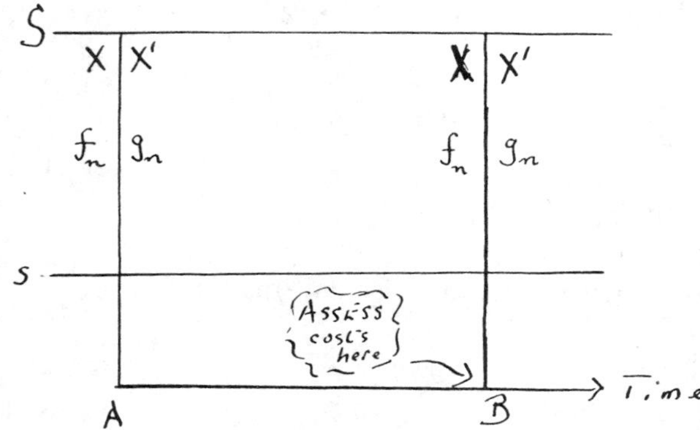

Let us assess operational costs in terms of X at B. First, no transport costs are payable unless $X<s$, and so their contribution to R, now a random variable, is $R=K$ with probability Σf_n taken over $-\infty<n<s$. Now suppose that there is no demand over AB. Then X at B$=X'$ at A and there is only the holding charge $X'Th$ component of R to consider. This is the case until X falls below zero. For instance, when $X=-1$ the holding charge and penalty charge are respectively

$$\frac{(\frac{1}{2}X')X'Th}{X'+1}, \text{ and } \frac{\frac{1}{2}Tp}{X'+1}.$$

The following table shows the level X at B, given X' at A, and the consequences of typical demand.

Demand	X	Holding Charge	Penalty Charge
0	X'	X'Th	0
1	X'-1	(X'-½)Th	0
2	X'-2	(X'-1)Th	0
3	X'-3	(X'-3/2)Th	0
.....	.....	.....	.....
X'	0	½X'Th	0
X'+1	-1	(½X')X'Th/(X'+1)	½Tp/(X'+1)
X'+2	-2	(½X')X'Th/(X'+2)	½.2.2Tp/(X'+2)
X'+3	-3	(½X')X'Th/(X'+3)	½.3.3Tp/(X'+3)
.....	.....	.....	.....

To complete the formulation of R thus requires knowledge of the distributions f_n and g_n of X and X'. To be precise, let

$$f_n=P[X=n] \quad (-\infty<n\leq S), \quad g_n=P[X'=n] \quad (s\leq n<S),$$
$$G_S=P[X'=S] \qquad (1)$$

The (s,S) procedure requires the following coupling between X' and X at A:

$$G_S = f_S + \sum_{n=-\infty}^{s-1} f_n$$

$$g_n = f_n \quad (n=s,s+1,\ldots,S-1). \qquad (2)$$

Also, connecting X at B with X' at A gives:

$$\begin{aligned} & f_S = G_S a_0 \\ & \ldots\ldots\ldots \\ & f_{S-n} = G_S a_n + g_{S-1} a_{n-1} + \ldots + g_{S-n} a_0 \qquad (n=1,2,\ldots,S-s) \\ & f_s = G_S a_{S-s} + g_{S-1} a_{S-s-1} + \ldots + g_s a_0 \\ & \ldots\ldots\ldots \\ & f_{s-n} = G_S a_{S-s+n} + \ldots + g_s a_n \qquad (n=1,2.....). \end{aligned} \qquad (3)$$

From these we can derive the following specific values which enable, in principle, the distribution of R, and hence r, to be obtained.

$$G_S = \frac{1}{1+(S-s)\bar{a}} \quad with\ \bar{a}=1-a.$$

$$g_{S-1} = g_{S-2} = \ldots = g_s = G_S \bar{a},$$

$$g_n = 0 \quad (n<s)$$

$$f_S = f_{S-1} = \ldots = f_s = G_S \bar{a},$$

$$f_{s-n} = G_S \bar{a} a^n, \qquad (n \geq 0).$$

The practical problem is to use these formulae to find advantageous values of s and S and, further, to study replenishment cycle time which, of course, entails making plausible assumptions about the connection between the demand distribution and time. See also [2] for further detail.

References

1. Beaumont, G.P. (1983) *Introductory applied probability*. Ellis Horwood. Chichester.

2. Prabhu, N.U. (1965) *Queues and inventories*. Wiley.

7

Mathematical Teasers

This chapter offers two mathematical teasers for amusement. A recent article [1] describes some variants on 7.1 which arrive at different answers.

7.1 PROBLEMS WITH PRODUCTS

Two friends, whom we shall call Mr. P and Mr. S (we are asked not to reveal their real names), were spending a pleasant evening at the local pub, when they saw Mr. Gamesmaster, the notorious bore, approaching. He asked them whether they would like to test their wits on a problem. "What sort of a problem?" groaned the two friends in chorus.

"Well, one with sums and products. If you are not interested we can discuss football results instead." Messrs. P. and S. opted for the problem.

"It is a very simple problem", Mr. G. began. "I am going to select two integers, m and n, say. Both are drawn from the range $2 \leq m \leq n \leq 99$. I shall write down on this piece of paper the sum $s = m + n$ and give it discreetly to Mr. S. Then I shall write down the product $p = mn$ on another piece of paper and hand it equally discreetly to Mr.P. Your problem is to find m and n by logical discussion - if you can." He then left them to their task.

After some time Mr. S. said: "I do not know what m amd n are, and neither do you." Mr.P. considered this and, after still more thought, said: "I know what m and n are." To which Mr. S replied: "And so do I."

So what are m and n?

Consider Mr.S.'s first statement. The first half is trivial. But what about the second half? Is it equally uninformative? Or is it imaginable that Mr.P. might already know something about m and n.

As a matter of fact, Mr. P might already know the two numbers, just by looking at the product he has been given. For instance, suppose the product had been 21. This can only be factorised into 3 times 7. Hence m must be 3 and n must be 7, and therefore s is 10. By asserting that Mr.P. can not know what Mr.S. holds, Mr.S. admits that he does not hold 10, because, if he did, he could not be sure that Mr. P. does not hold 21 and thus be able to identify m and n. Generally, if Mr.S. were to hold a number which can be partitioned into two elements, both of which are primes, then the product of these two integers would give away their identity. Thus, by the second half of his first statement Mr.S. has disclosed that he does not have such a number. He has informed Mr. P. that he, Mr.S., does not hold 4=2+2, 5=2+3, 6=3+3, 7=5+2, 8=3+5, 9=2+7, 10=3+7, and so on. On the other hand, Mr.S. could, as far as this argument goes,hold 11, because none of 2+9,3+8,4+7,5+6 is a partition into two primes.This excludes quite a number of (m,n) couples, but Mr.P. wants to reduce the possibilities to just one single couple. He looks therefore for further exclusions. He remembers something called the **Goldbach Conjecture**, a claim that every even integer can be expressed as the sum of two primes in at least one way. The conjecture is still unproved after 250 years, but for integers up to 99, the maximum possible according to the rules of this game, its truth can be verified by manageable computation. Thus Mr.S. can not have an even sum in his hand.

Intrigued by these thoughts, Mr.P. decides to concentrate on the possible odd integers that Mr.S. might hold from 11 onwards. This results in the following list.

Possible s	**Possible p=mn**
11	18,24,28,30
17	30,42,52,60,66,70,72
23	42,60,76,90,102,112,120,126,130,132
27	50,72,92,110,126,140,152,162,170,176,180,182
29	54,78,100,120,138,154,168,180,190,198,204,208,210
35	66,96,124,150,174,196,216,234,250,264,276,286,294, 300,304,306
37	70,102,132,160,186,210,232,252,270,286,300,312,322, 330,336,340,342
41	78,114,148,180,210,238,264,288,310,330,348,364,378, 390,400,408,414,418,420
47	90,132,172,210,246,280,312,342,370,396,420,442,462, 480,496,510,522,532,540,546,550,552
51	98,144,188,230,270,308,344,378,410,440,468,494,518, 540,560,578,594,608,620,630,638,644,648,650
53	102,150,196,240,282,322,360,396,430,462,492,520,546, 570,592,612,630,646,660,672,682,690,696,700,702
....	

Looking carefully at this table Mr. P observes that many of the possible products p are duplicates. For example, 30 occurs opposite s=11 and 17. The duplicates worry him. In his excitement he has not yet looked at the number he himself holds, but he realises that if this number were, for instance, 30, then he would not know whether s equals 11 or 17. Therefore he sets about constructing a new table in which the duplicates are eliminated. He obtains the following:

11 18,24,28
17 52
23 76,112,130
27 50,92,110,140,152,162,170,176,182
29 54,100,138,154,168,190,198,204,208
35 96,124,174,216,234,250,276,294,304,306
37 160,186,232,252,336,340
41 114,148,180,238,288,310,348,364,390,400,408,414,418
47 132,172,246,280,370,442,480,496,510,522,532,550,552
51 98,144,188,230,308,344,410,440,468,494,518,560,578,
594,608,620,638,644,648,650
53 102,240,282,360,430,492,520,570,592,612,646,660,672,
682,690,696,700,702

"How curious", he exclaims. Corresponding to the possible sum 17 there is the unique product 52. I wonder what is on my paper". He looks, sees that he has 52, realises that m=4 and n=13, and tells Mr.S that he now knows what the values of m and n are without telling him their values.

Mr.S has been doing the same arithmetic and, when he hears that Mr.P. knows the answer and realises that he himself holds 17, he knows the answer as well.

We have called Mr.G. a bore. But we see that he knows about numbers. He chose m and n in such a way that the puzzle can be solved. This would not have been the case if he had restricted the range of possibilities to numbers not exceeding 15. Then the line 17 of the second list would contain 66 as well as 52, and though Mr.P. knows that he himself holds 52, and can therefore guess that s=17, S would not be able to determine whether P has 52 or 66, even though he knows that his own number is 17.

Imagine now that the limits for m and n are not 99, but some larger number. What does the puzzle then look like? We must leave that to the contemplation of the reader. To paraphrase a famous mathematician of the 17-th century, our pages are too narrow to extend our search much further.

Those readers with a taste for computer programming may care to automate the arithmetic performed by the players.

7.2 A PUZZLE SERIES

One day, a colleague whose research interest lay in the area of statistical mechanics, told me that, by physical intuition, he had arrived at a result that he was unable to prove. It had seemed so improbable that he had fallen back on numerical computation, and this had confirmed his suspicions. I was very busy preparing remedial lectures on elementary probability, but sufficiently intrigued to enquire further. This is essentially what he told me.

Let x_0, y_0 be real numbers. Then

$$y_0 + \ln\{\cosh x_0 + (\sinh^2 x_0 + e^{-4y_0})^{1/2}\} = \frac{1}{2}\sum_{n\geq 0} \frac{\psi(x_n, y_n)}{2^n} \qquad (7.2.1)$$

where

$$\psi(x,y) = \frac{\ln\ \cosh\ x\ + \ln\ 4B(x,y)}{2}, \qquad (7.2.2),$$

$$B^2(x,y) = \frac{\cosh\ 4y\ +\ \cosh\ 2x}{2} \qquad (7.2.3)$$

and x_n, y_n are given by the recurrences

$$\exp(x_{n+1}) = \frac{B(x_n, y_n)}{\cosh(2y_n - x_n)} \exp(x_n), \qquad (7.2.4)$$

$$\exp(2y_{n+1}) = \frac{B(x_n, y_n)}{\cosh\ x_n}. \qquad (7.2.5)$$

"Ci vorrebbe un Ramanujan per dimostrare questo", exclaimed an Italian research student who happened to be standing nearby, meaning that Ramanujan, in his opinion, would have found the problem an interesting one. "Nevertheless", I said, "without a séance it will be hard to get his advice, so let us first see what we can do on our own. As it is always useful to allocate a letter to something let us call the left hand side of (1) S_0 in recognition of the subscript 0. If we then assume that S_0 can be expressed as the sum on the right I think you will agree that

$$2S_0 = \psi_0 + \psi_1/2 + \psi_2/2^2 + ..., \quad S_1 = \psi_1/2 + \psi_2/2^2 + ...,$$

giving

$$\psi_0 = 2S_0 - S_1, \textit{ or, replacing the variables,}$$

$$\psi(x_0, y_0) = 2S(x_0) - S(x_1, y_1). \qquad (7.2.6)$$

The problem can then be restated in equivalent form:

Given the expression S_0, the series representation but not the explicit form of ψ, and the recurrences (7.2.4) and (7.2.5), find ψ_0."

My companions seemed to agree.

Then I suggested, with a flash of inspiration, that we define two new quantities A_0, u_0 by

$$\cosh x_0 = A_0 \cosh u_0$$
$$(\sinh^2 x_0 + e^{-4y_0})^{1/2} = A_0 \sinh u_0 \qquad (7.2.7)$$

so that

$$A_0^2 = 1 - e^{-4y_0}, \textit{ and}$$
$$S_0 = y_0 + u_0 + \ln A_0. \qquad (7.2.8)$$

"We shall then be in business", I exclaimed excitedly, "if we can prove the following two Lemmata:

Lemma 1:

$$A_1 = A(x_1, y_1) = \frac{\sinh 2y_0}{B_0}. \qquad (7.2.9)$$

Lemma 2:

$$u_1 = 2u_0. \qquad (7.2.10)$$

Given these, we can perform the following calculation:

$$e^{\psi_0} = \frac{e^{2S_0}}{e^{S_1}} = \frac{e^{2y_0+2u_0}A_0^2}{e^{y_1+u_1}A_1} = \frac{A_0^2 e^{2y_0}}{A_1 e^{y_1}}$$
$$= \frac{A_0^2 e^{2y_0} B_0}{e^{y_1}\sinh 2y_0} = \frac{A_0^2 e^{2y_0} B_0 (\cosh x_0)^{1/2}}{\sinh 2y_0 B_0^{1/2}}$$

(*this uses* (7.2.5)),

and so, finally,

$$e^{\psi_0} = \frac{e^{2y_0}(1-e^{-4y_0})(B_0\cosh x_0)^{1/2}}{\sinh 2y_0} = (4B_0\cosh x_0)^{1/2},$$

which is the exponential form of (7.2.2). The job is done, apart from proving the Lemmata."
Here are proofs.

Proof of Lemma 1.

$$A_1^2 = 1-e^{-4y_1} = 1-\frac{\cosh^2 x_0}{B_0^2}$$

$$= \frac{\frac{1}{2}(\cosh 4y_0+\cosh 2x_0)-\cosh^2 x_0}{B_0^2} \quad \textit{(by (7.2.8) and (7.2.3)},$$

and this reduces to $(\sinh 2y_0/B_0)^2$, and completes the proof of the Lemma.

Proof of Lemma 2.

We show that $\cosh u_1=\cosh 2u_0$. In a similar manner we can show that $\sinh u_1=\sinh 2u_0$ and the Lemma follows.

By (7.2.7) we have $\cosh u_1=(\cosh x_1)/A_1$. Then

$$2\cosh x_1 = e^{x_1}+e^{-x_1}$$

$$= \frac{B_0 e^{x_0}}{\cosh(2y_0-x_0)} + \frac{\cosh(2y_0-x_0)}{B_0 e^{x_0}}$$

$$= \frac{B_0^2 e^{2x_0}+\cosh^2(2y_0-x_0)}{B_0 e^{x_0}\cosh(2y_0-x_0)}.$$

The numerator can be reduced to

$$\cosh(2y_0-x_0)\{e^{2x_0}\cosh(2y_0+x_0)+\cosh(2y_0-x_0)\}$$

and, after some more manipulation we get

$$2\cosh x_1 = \frac{A_0^2 e^{2y_0} \cosh 2u_0}{B_0}$$

from which follows

$$\cosh u_1 = \frac{\cosh x_1}{A_1} = \frac{\sinh 2y_0 \cosh 2u_0}{B_0 A_1} = \frac{\sinh 2y_0 \cosh 2u_0}{\sinh 2y_0}$$

by Lemma 1, and this completes the proof of Lemma 2.

"I expect you will need to check this" I said modestly to my small audience, "and I should be glad if you could find a less tortuous proof".

As presented, this problem has the form: Given S_0 and the recurrences, find ψ. This is a practical problem too. It *might* be computationally efficient to evaluate S_0 indirectly by using such a series, and what would be the restrictions on the recurrences? I have explored the question with other functions S and the **same** recurrences. Can other recurrences be found which improve the efficiency of this proposal? And finally, a theoretical problem: Given ψ and the recurrences, find S. There seems scope for some pleasurable exercises in this field.

Reference

1. Sallows, L. (1995) The impossible problem. Mathematical Intelligencer. 17, 27-33.

8

Triangular Geometry

Study 8.1 is concerned primarily with the geometry of a rarely discussed point in a triangle having the minimizing property referred to in Chapter 4. The points and circles related to triangles have a huge literature. A recent reference is [4], while [7] discusses the influence of Napoleon Buonaparte. (See also 8.2.5). 8.2 is a discursive study of some pedal properties of triangles which have intrigued many mathematicians including Hilbert. This fascinating subject is also resurrected from time to time: a recent article is [3].

8.1 FERMAT, TORRICELLI ET AL.

8.1.1 The Torricelli Point

Let a triangle $A_0B_0C_0$ be given, and let A_1, B_1, C_1, all three outside the triangle, be the vertices of three equilateral triangles with bases B_0C_0, C_0A_0, A_0B_0, respectively. We distinguish between the two cases where none of the angles of $A_0B_0C_0$ exceeds 120° (Case (a),Figure 8.1.1), and where one of them does (Case (b),Figure 8.1.2).The case where one angle is equal to 120^0 is an easily recognizable frontier between these two cases. Observe that the intersection of A_0A_1 and B_0B_1, denoted by T, lies in Figure 8.1.1 inside, and in Figure 8.1.2 outside, the original triangle. To start we prove the

Lemma: The triangles $A_0C_0A_1$ and $B_1C_0B_0$ are congruent.

<u>Proof:</u> $C_0B_1 = C_0A_0$ and $C_0B_0 = C_0A_1$.
Moreover, in case (a), $\angle B_1C_0B_0 = \angle A_0C_0B_0 + 60^o = \angle A_0C_0A_1$,
and, in case (b), $\angle B_1C_0B_0 = \angle A_0C_0B_1 + 60^o = \angle A_0C_0A_1$.
This completes the proof of the Lemma.

Denote $\angle C_0A_0A_1 = \angle C_0A_0T$ by θ and $\angle C_0B_0B_1 = \angle C_0B_0T$ by ϕ.
It follows from the Lemma that also

$$\angle C_0B_1B_0 = \angle C_0B_1T = \theta$$

and $$\angle C_0A_1A_0 = \angle C_0A_1T = \phi.$$

Hence $C_0B_1A_0T$ and $C_0A_1B_0T$ are cyclic quadrilaterals in case (a), and $C_0A_0B_1T$ and $C_0B_0A_1T$ are cyclic quadrilaterals in case (b). In Figure 8.1.1 we have $\angle A_0TC_0$ = $\angle B_0TC_0$ =120°; hence also $\angle A_0TB_0$=120°, so that $A_0TB_0C_1$ is a third cyclic quadrilateral. Similarly, in Figure 8.1.2 we have $\angle A_0TC_0$ = $\angle B_0TC_0$ = 60°, and A_0TB_0 = 120°, so that $A_0TB_0C_1$ is a third cyclic quadrilateral as well. Observe that in either case all three quadrilaterals pass through T, the point defined as the intersection of A_0A_1 and B_0B_1.

Figure 8.1.1 **Figure 8.1.2**

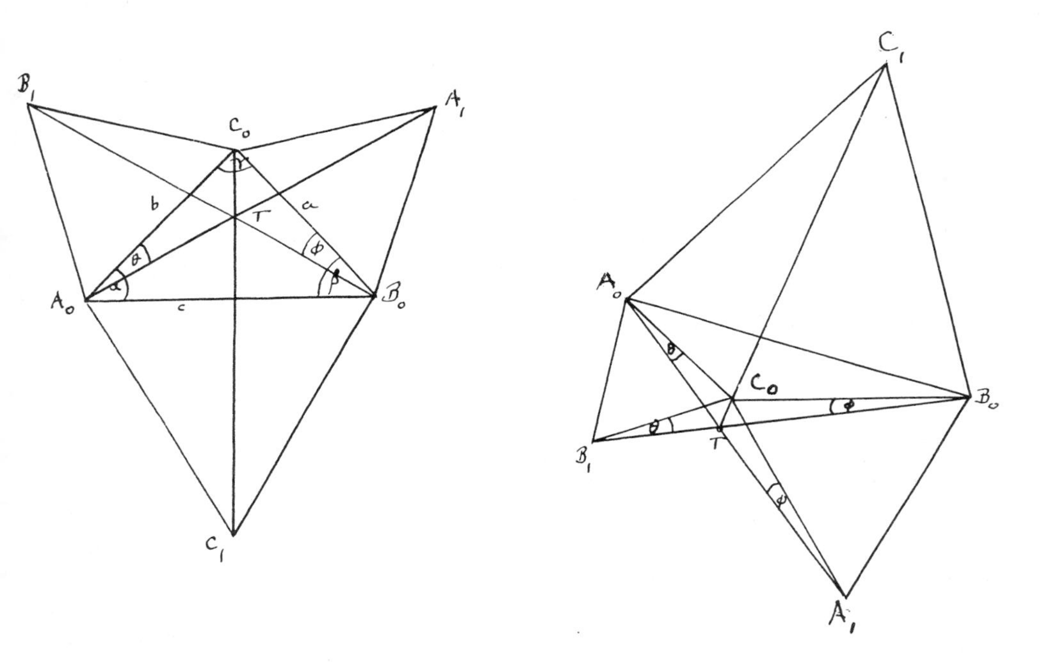

We now prove

Theorem 1: The straight line C_0C_1 passes also through T both in cases (a) and (b).

Proof: (a) Because $A_0TB_0C_1$ is cyclic, $\angle A_0C_0C_1$ = $\angle A_0B_0C_1$ = 60°. We know already that $\angle A_0TC_0$ = 120°, and so C_0TC_1 is a straight path.

(b) $\angle A_0TC_1$ = 60°. We know already that $\angle A_0TC_0$ = 60°; hence TC_0C_1 is a straight path.

The main result of this exercise is Theorem 2, which is expressed in two forms.

Theorem 2a: Let $A_0B_0C_0$ be a triangle, no angle of which exceeds 120°. Then the point D=T minimises the sum of distances $DA_0+DB_0+DC_0$. T is shown in Figure 8.1.1.

Figure 8.1.3

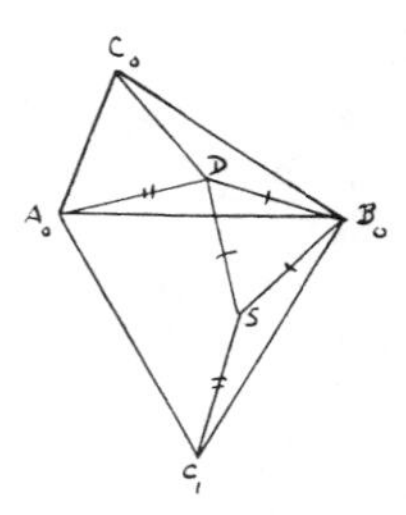

Proof: Consider Figure 8.1.3. D is a point internal to triangle $A_0B_0C_0$. Construct the equilateral triangle $A_0B_0C_1$, and the equilateral triangle SB_0D, with S nearer to C_1 than to C_0. It follows easily that the triangles SB_0C_1 and DB_0A_0 are congruent, so that the sum of the three straight line segments $DA_0+DB_0+DC_0$ is equal to the length of the path C_1SDC_0. If now for D we choose point T, then since $\angle TC_1B_0 = \angle TC_1A_0 = 60°$, both T and S will lie on the straight line C_0C_1; the path C_0TSC_1 will have minimum length, and

$$TA_0 + TB_0 + TC_0 = C_0C_1.$$

By an analogous argument the smallest sum above is equal also to A_0A_1 and B_0B_1, and these three segments have equal lengths.

The lines A_0A_1, B_0B_1, C_0C_1 have been called Simpson Lines after Thomas Simpson (1710-1761). He is not to be confused with Simson, mentioned in 8.2.

Theorem 2b: Let $A_0B_0C_0$ be a triangle with $\angle A_0C_0B_0$ larger than 120°. Then C_0=D minimises $DA_0+DB_0+DC_0$. T is shown in Figure 8.1.1.
For the proof the reader is referred to [5].

Recall that in case (a) we had $TA_0+TB_0+TC_0=C_0C_1$. In the present case (b), when the largest angle is at vertex C_0, we have

Theorem 3: $TA_0 + TB_0 - TC_0 = C_0C_1$.

Proof: Refer to Figure 8.1.4. The triangle A_0TB_0 has, at T, an angle of 120°, so that Theorem 2a applies to it, thus:

$$TA_0 + TB_0 + TT = TC_1.$$

Since TT=0, we have

$$TA_0 + TB_0 = TC_1 = TC_0 + C_0C_1.$$ **QED**

Figure 8.1.4

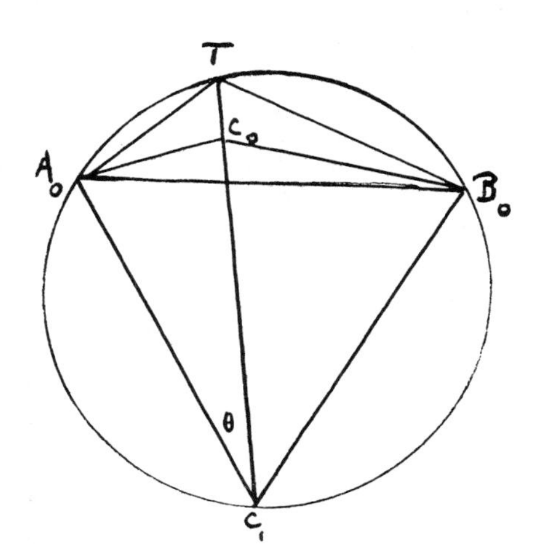

In [5] it is stated that in this case the point D=T minimizes $DA_0+DB_0-DC_0$. This is an error which we understand ([6]) will be corrected by a future publication.

Historical Note: Early in the seventeenth century P. Fermat posed the following problem. Given three points in the plane, find the point such that the sum of its distances from the three given points is a minimum. E. Torricelli proved that if no angle of the triangle formed by the given points exceeds 120°, then it is necessary and sufficient for the point sought that the angles which the straight lines to the given point form are all 120°. He constructed the required point as the intersection of two circles. The solution to case (b) is mentioned without proof.

Although algebraic methods are often deprecated we shall nevertheless discuss briefly their use in finding the point P that minimizes the distance S=PA+PB+PC relative to an arbitrary triangle ABC. Figure 8.1.5 shows a triangle in which no angle exceeds 120° ; the side lengths are AB=c, BC=a, CA=b. The angle at A is denoted by α. With A as origin and AB as axis the point P with polar coordinates (r,θ) is defined.

Figure 8.1.5

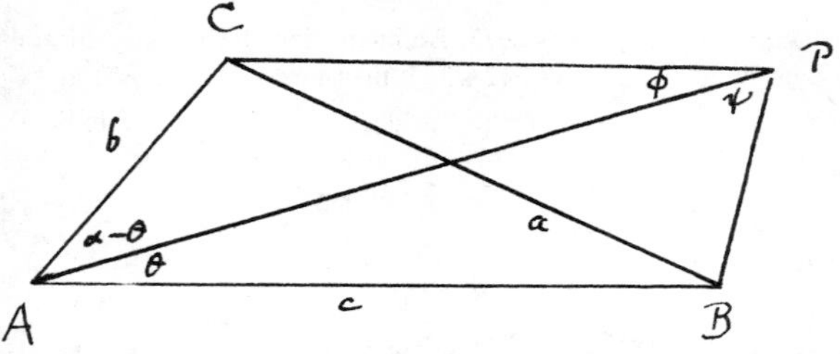

As shown, the point P is external to the triangle. Denote the angles APC and BPA by ϕ and ψ, respectively. Whatever θ, moving P towards A plainly diminishes S. and it is clear that no point P external to ABC can minimize S. In Figure 8.1.5 we see that holding θ constant and reducing r increases ϕ and ψ which become such that their sum equals π when P meets BC. When P enters the triangle the sum exceeds π.To obtain an expression for S we use the cosine rule to give

$$PC^2 = b^2 + r^2 - 2br\cos(\alpha - \theta),\quad PB^2 = c^2 + r^2 - 2rc\cos\theta.$$

Then

$$\frac{\partial PB}{\partial r} = \frac{r - c\cos\theta}{PB} = \cos\psi,\quad \frac{\partial PC}{\partial r} = \frac{r - b\cos(\alpha - \theta)}{PC} = \cos\phi,$$

$$\frac{\partial PB}{\partial \theta} = \frac{cr\sin\theta}{PB} = r\sin\psi,\quad \frac{\partial PC}{\partial \theta} = \frac{-br\sin(\alpha - \theta)}{PC} = -r\sin\phi,$$

by the sine rule.

For P to minimize S, r and θ must simultaneously satisfy $\partial S/\partial r = \partial S/\partial \theta = 0$, that is

$$1 + \cos\psi + \cos\phi = 0,\quad r(\sin\psi - \sin\phi) = 0.$$

This holds for maximization too, but there is no danger of confusion here.
A solution is $\phi = \psi = 2\pi/3$ and, as long as no angle of ABC exceeds $2\pi/3$, this means that P is the internal Torricelli point T. The values of r and θ are given by

$$r = \frac{bc\sin(2\pi/3 - \alpha)}{\Delta^{1/2}\sin\pi/3},\quad \tan\theta = \frac{n}{d},$$

$$n = b\sin(\alpha - \frac{\pi}{3}) + c\sin\frac{\pi}{3},\quad d = b\cos(\alpha - \frac{\pi}{3}) + c\cos\frac{\pi}{3},$$

$$\Delta = n^2 + d^2.$$

These formulae are useful for calculating the position of the Torricelli point which otherwise can be awkward.

As stated above, whatever the angles of the triangle, a point P external to the triangle can not minimize S. When the angle $\alpha = 2\pi/3$, the Torricelli point coincides with the vertex A, and when $\alpha > 2\pi/3$, it lies outside the triangle, in which case it can not minimize S. The minimizing point in this case can neither lie inside nor outside the triangle: it has therefore to lie on the boundary and a moment's thought shows that it has to be the vertex where the angle exceeds $2\pi/3$. An interesting check on the formulae is to find α so that the angle at one of the other vertices, C, say, is $2\pi/3$. It then turns out that $r=b$ and $\theta=\alpha$.

Another interesting exercise is to identify the locus of the minimizing point as α varies. It turns out to be the circle with centre at $(c/2, -c/2 \cot \pi/3)$ (in cartesian coordinates relative to A as origin and AB as x-axis), and radius $c(\operatorname{cosec}\pi/3)/2$. This passes through A ***and is independent of b***.

8.1.2 Nests of Torricelli Triangles

The basic construction of $A_1B_1C_1$ on $A_0B_0C_0$ as base by equilateral triangles can be repeated using $A_1B_1C_1$ as base to construct in similar fashion a triangle $A_2B_2C_2$. And this process can be repeated indefinitely. We show

Theorem 1 $C_2C_0 = C_0C_1$.

This is the equivalent of $C_2C_1 = 2C_0C_1$. The result holds whether or not T is an interior point of the base triangle.

Proof Figure 8.1.6 may be consulted for reference.

Figure 8.1.6

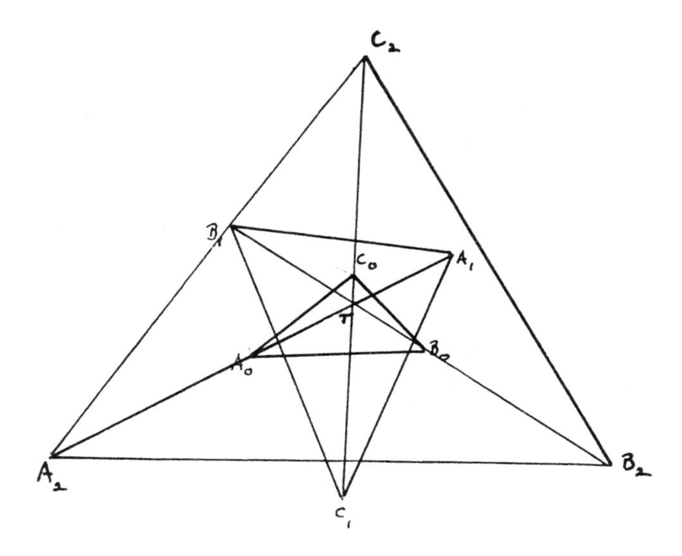

Just as in **8.1.1**, Theorem 1, we concluded that C_0TC_1 is a straight line, so also we can conclude that C_1TC_2 is a straight line and hence that C_1,T,C_0,C_2 are collinear. Given this, we have, applying Theorem 2a to the triangle $A_1B_1C_1$,

$$C_2C_1 = TA_1 + TB_1 + TC_1 = A_0A_1 - TA_0 + B_0B_1 - TB_0 + C_0C_1 - TC_0$$

$$= 3C_0C_1 - C_0C_1 = 2C_0C_1.$$

Thus

$$C_0C_1 = C_0C_2.$$

Similarly,

$$A_0A_1 = A_0A_2,\ B_0B_1 = B_0B_2.$$ **QED**

Figure 8.1.7 is an instance where T is exterior to $A_0B_0C_0$, but interior to $A_1B_1C_1$.

Figure 8.1.7

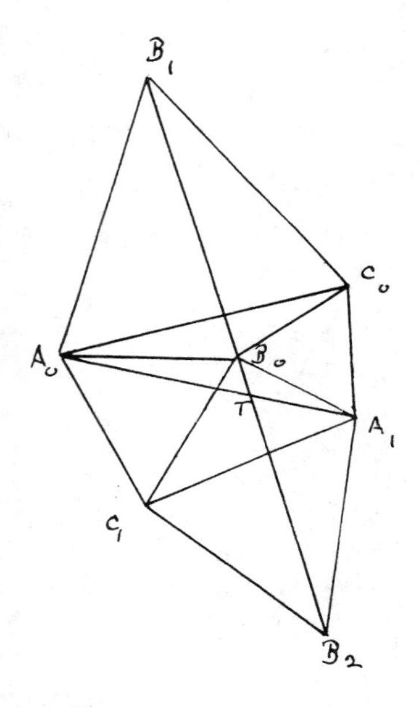

Since T is interior to triangle $A_1B_1C_1$, by Theorem 2a,

$$B_1B_2 = TA_1 + TB_1 + TC_1 = A_0A_1 - TA_0 + B_0B_1 + TB_0 + C_0C_1 - TC_0$$

$$= 3B_0B_1 - (TA_0 + TC_0) + TB_0$$

$$= 3B_0B_1 - (B_0B_1 + 2B_0T - B_0T) + B_0T$$
$$= 2B_0B_1.$$ **QED**

Theorem 1 shows how a sequence of triangles $A_nB_nC_n$ can be constructed using the "equilateral triangle method". We can start with $A_0B_0C_0$, construct $A_1B_1C_1$, and infer $A_nB_nC_n$ for n=2,3,.... The coordinates of the vertices form a sequence. It suffices to investigate the sequence (A_n).

Let A_0A_1 be a coordinate axis such that A_n has coordinate x_n. Then, by the Theorem,

$$x_2=x_1-2(x_1-x_0)=2x_0-x_1$$

$$x_3=x_2-2(x_2-x_1)=2x_1-x_2$$

$$x_4=x_3-2(x_3-x_2)=2x_2-x_3$$

....

$$x_{n+2}=2x_n-x_{n+1},$$

giving

$$x_{n+2}+x_{n+1}-2x_n=0.$$

Taking $x_0=0$, $x_1=X$, so that distances are measured from A_0 and the unit of length is $X=A_0A_1$, we obtain by the standard method,

$$x_n = \tfrac{1}{3}X\{1-(-2)^n\}.$$

As the same argument holds for the sequences (B_n), (C_n) of vertices we can construct a sequence of triangles all having the same *Torricelli Point* T and enjoying the properties we have found for the initial triangles. This is Theorem 2.

Theorem 2 A sequence of triangles $(A_nB_nC_n)$ can be constructed using the rule $x_n=\tfrac{1}{3}X\{1-(-2)^n\}$ for the coordinates of the vertices A_n, B_n, C_n relative to origins A_0, B_0, C_0 using the common unit of length $X=A_0A_1=B_0B_1=C_0C_1$. The triangles have the common *Torricelli Point* T and, among other shared properties, the analogous cyclic quadrilaterals of which $C_0TA_0B_1$ in Figure 8.1.1 is an example.

The sequences of vertices can be continued also for negative integer n. Thus, the triangle $A_{-1}B_{-1}C_{-1}$ is such that the equilateral triangle construction on it as base triangle yields triangle $A_0B_0C_0$. In the limit as $n \to -\infty$, a triangle $A_fB_fC_f$ is reached such that $A_0A_f=B_0B_f=C_0C_f=\tfrac{1}{3}X$, where X is the common unit defined in Theorem 2. Figure 8.1.8 illustrates the limiting triangle. We state

Theorem 3 The limiting triangle $A_fB_fC_f$ is equilateral and retains, among others, the property that $C_fT=A_fT=B_fT$.

Proof By Theorem 2, the cyclic quadrilateral property is retained by $C_fTA_fB_f$, the

B-vertices having in the limit merged. Since angle $A_fTC_f = 2\pi/3$, the opposite angle $A_fB_fC_f = \pi/3$. Also angle $A_fTB_f = \pi/3 =$ angle $A_fC_fB_f$. Hence triangle $A_fB_fC_f$ is equilateral. The remaining property is proved by argument similar to that used for Theorem 2.

Figure 8.1.8

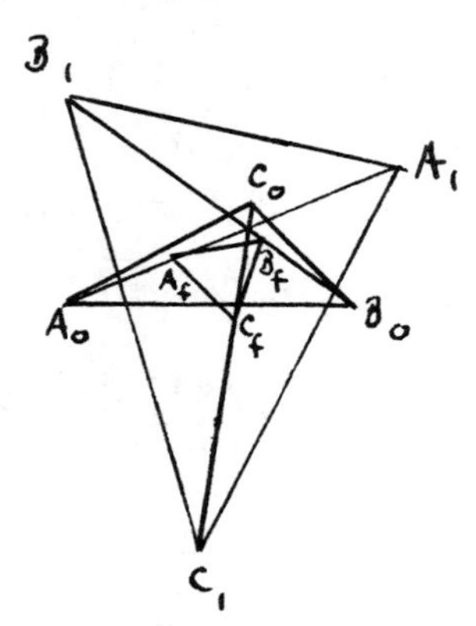

When $A_0B_0C_0$ is equilateral, the Torricelli point T is the common centre of the in-circle of $A_0B_0C_0$ and of its circumscribed circle. It is easily checked that $A_0T = \frac{1}{3}A_0A_1$, and likewise that B_0T, C_0T have the same length. T is thus the *confluence* of the points $A_fB_fC_f$, and in the limit the triangle $A_fB_fC_f$ reduces to the *point triangle* T. Construction of the sequence $A_{-n}B_{-n}C_{-n}$ of interior triangles forms a pretty pattern worthy of anyone's bathroom floor.

This discussion would be incomplete without reference to the so-called Napoleon's Theorem. If in Figure 8.1.1 the centres of the three external equilateral triangles are joined Napoleon Buonaparte is alleged to have claimed to Lagrange that the resulting triangle is equilateral. For a proof see the Notes at the end of this Chapter. The Theorem is supposed to be true when *internal* equilateral triangles are inscribed on the sides of the base triangle. See Wetzel [7].

8.2 PEDALMANIA

8.2.1 Pedal circles and the Simson line

Reference is made to Figure 8.2.1.

Consider a triangle ABC and a point P in the same plane. Define the points X,Y,Z as the feet of the perpendiculars from P to the sides AB,BC,CA, respectively. The circle through X,Y,Z is called the pedal circle of P with respect to the triangle ABC. It is the circumcircle of the triangle XYZ, which might be called the pedal triangle of P with respect to the triangle ABC.In particular, if P is the incentre of ABC, the pedal circle is the incircle. If P is the circumcentre, or the orthocentre, of ABC, the pedal circle is the nine-point circle of ABC. Some reminders about the nine-point circle are given in the Notes. Choose the centre O of the circumcircle as origin of (x,y) coordinates. Let the radius of the circumcircle be R, and denote the angles xOA,xOB,xOC, measured positively anticlockwise, by α,β and γ. The coordinates of A are thus ($R\cos\alpha,R\sin\alpha$), and those of B and C are obtained respectively by replacing α by β and γ. This system was suggested by Jones in the venerable text-book [1] and has the advantage of allowing symmetry to substitute for much calculation. Some details to help the following descriptions are given in the Notes.

Figure 8.2.1

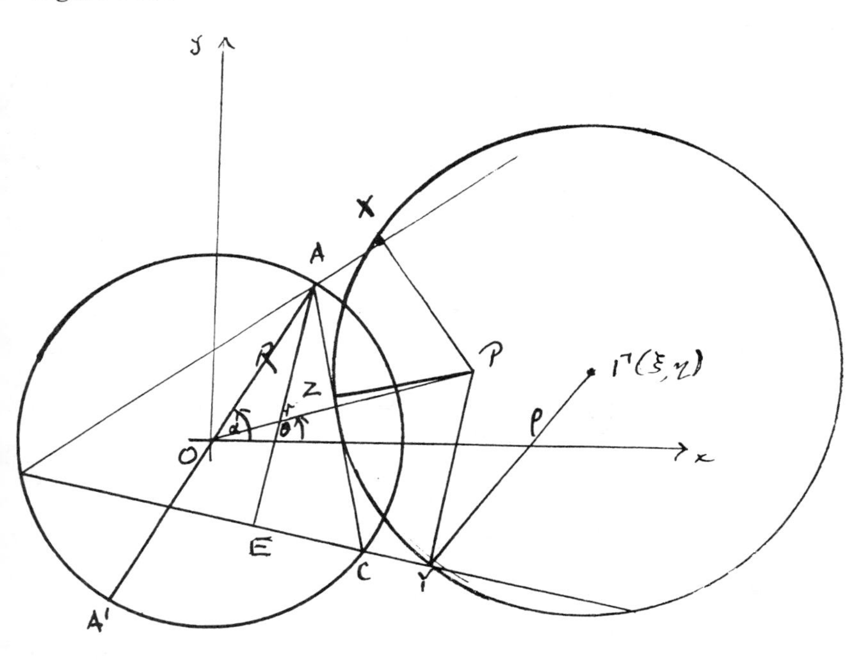

Let P be the point ($r\cos\theta,r\sin\theta$), and denote the centre of the pedal circle by $\Gamma(\xi,\eta)$, and its radius by ρ. Then,

$$\xi=\frac{\lambda_1 r^3-\lambda_2 r^2 R+\lambda_3 r R^2-\lambda_4 R^3}{2(r^2-R^2)},$$

$$\eta=\frac{\mu_1 r^3-\mu_2 r^2 R+\mu_3 r R^2-\mu_4 R^3}{2(r^2-R^2)},$$

$$\rho=\left|\frac{\{r^6-2\nu_1 r^5 R+\nu_2 r^4 R^2-2\nu_3 r^3 R^3+\nu_2 r^2 R^4-2\nu_1 r R^5+R^6\}^{1/2}}{2(r^2-R^2)}\right|,$$

where

$$\lambda_1=\cos\theta,\quad \lambda_2=\cos(\alpha+\beta+\gamma-2\theta),$$

$$\lambda_3=\cos(\alpha+\beta-\theta)+\cos(\beta+\gamma-\theta)+\cos(\gamma+\alpha-\theta),$$

$$\lambda_4=\cos\alpha+\cos\beta+\cos\gamma,$$

$$\mu_1=\sin\theta,\quad \mu_2=\sin(\alpha+\beta+\gamma-2\theta),$$

$$\mu_3=\sin(\alpha+\beta-\theta)+\sin(\beta+\gamma-\theta)+\sin(\gamma+\alpha-\theta),$$

$$\mu_4=\sin\alpha+\sin\beta+\sin\gamma.$$

(8.2.1a)

The equation of the pedal circle is

$$(x-\xi)^2+(y-\eta)^2=\rho^2. \qquad (8.2.1b)$$

When P lies on a diameter, special forms for ξ,η and ρ are obtained. On the segment OA of the diameter A'A

$$\xi=\frac{r^2\cos\alpha+rR\{\cos\alpha-\cos(\beta+\gamma-\alpha)\}+R^2(\cos\alpha+\cos\beta+\cos\gamma)}{2(R+r)},$$

η =*ditto with sines,*

$$2(R+r)\rho=[r^4-2r^3R\{\cos(\alpha-\beta)+\cos(\gamma-\alpha)\}+$$

$$+2r^2R^2\{1+\cos(\beta-\gamma)+\cos(\beta+\gamma-2\alpha)\}-2rR^3\{\cos(\alpha-\beta)+\cos(\gamma-\alpha)\}+R^4]^{1/2}.$$

(8.2.1c)

For points P on A'O, replace θ by $\alpha+\pi$, or simply replace r by -r in (8.2.1c). For P on diameters B'B, C'C, symmetry provides the appropriate equations.

Theorem 1 The circles defined by (1c) pass through the foot of the altitude from A to BC. See Figure 8.2.2. An algebraic proof is given in the Notes. Here is a Euclidean proof. See Figure 8.2.3.

Figure 8.2.2

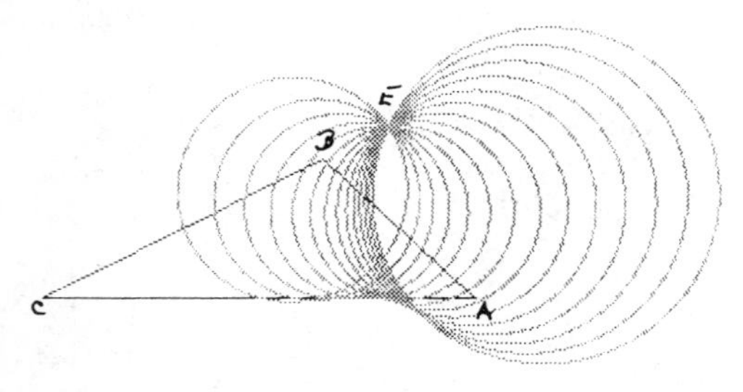

Figure 8.2.3

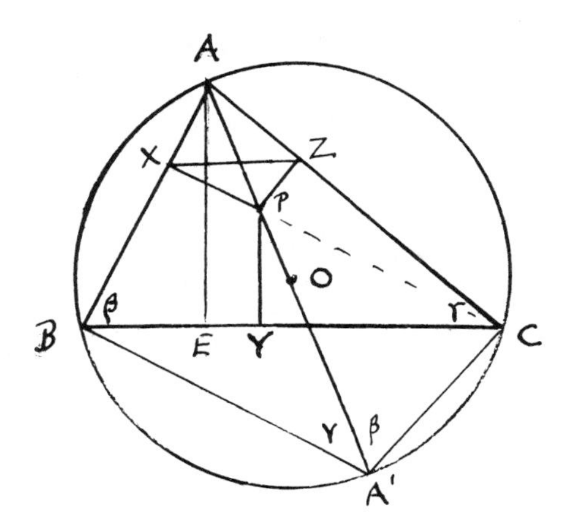

ABC is a triangle with circumcentre O. The circumcircle is shown in the Figure. AA' is the diameter starting at vertex A, and P is a point on it, interior to the circle, though this is not a restriction on the result. Angles ACA' and ABA' are right angles since they are angles in the semicircles. The angles in the triangles ABC at B and C are denoted by β and γ, respectively. Since ABA'C is a cyclic quadrilateral with right angles at at B and C it follows that the angles shown at A' are also γ and β, and since angle ACA' is right, that angle CAA' is $\pi/2$ - β.From the point P, perpendiculars to the sides of the triangle are drawn, meeting the sides at X,Y,Z, as shown. The pedal circle is the circle through X, Y, and Z. Since PZ and A'C are perpendicular to AC they are parallel. Thus angle CA'A=angle ZPA=β. Moreover, since AXPZ is a cyclic quadrilateral, β=angle ZPA=angle ZXA, and by similar reasoning angle AZX = γ. Thus, XZ is parallel to BC.

Now construct the altitude from A and let it meet XZ at E' and BC at E. Since XZ is parallel to BC, angle AE'X is right.
Right-angled triangles AXE' and ABE are similar, the angle at A also being $\pi/2$ - β. Also APZ and AA'C are similar right-angled triangles, and each of these is similar to AXE' and ABE.
It follows that

$$\frac{XE'}{PZ}=\frac{AE'}{AZ}=\frac{AE}{AC}=\frac{AE-AE'}{AC-AZ}=\frac{EE'}{ZC},$$

and hence that the right-angled triangles XE'E and PZC are similar. Therefore, the angles XEE' and PCZ are equal.
But angle PCZ = angle PYZ since PZCY is a cyclic quadrilateral. Hence XZYE is a cyclic quadrilateral, for the sum of the angles at X and Y is

E'XE+ZYE=$\pi/2$ - angle XEE' + $\pi/2$ + angle PYZ = π.

This proves Theorem 8.2.1.

While there exists a pedal circle when r=R and $\theta=\alpha$, that is, when P is A, that is not the case when P is A'. Then, the denominator of (8.2.1c), containing r-R, tends to zero as r->R, and the pedal circle turns into a straight line. That this straight line is BC can be confirmed by the property that the angles in a semicircle are right.

That in general, when P lies on the circumcircle ABC, pedal circles degenerate into straight (pedal) lines, follows from the occurrence of r^2-R^2 in (8.2.1a). This line is shown in the Notes to have equation

$$\{x-\frac{R}{2}(\cos\theta+\cos\alpha+\cos\beta+\cos\gamma)\}\sin\frac{1}{2}(\alpha+\beta+\gamma-\theta)=$$

$$=\{y-\frac{R}{2}(\sin\theta+\sin\alpha+\sin\beta+\sin\gamma)\}\cos\frac{1}{2}(\alpha+\beta+\gamma-\theta).$$

(8.2.2)

But the points A,B,C, have a dual character, possessing both a pedal circle and a pedal line. The pedal lines are the altitudes drawn from A,B,C. For instance, if P is A, the points X and Z are confluent at A, while Y becomes E.

The straight line (8.2.2) has been called the Pedal Line, Simson Line and Wallace Line. Robert Simson (1687-1768) was Professor of Mathematics in the University of Glasgow from 1711 to 1761, and W. Wallace named the line after Simson in 1772.

It is of interest to prove the pedal line property by purely geometric means. Reference is made to Figure 8.2.4.

Figure 8.2.4

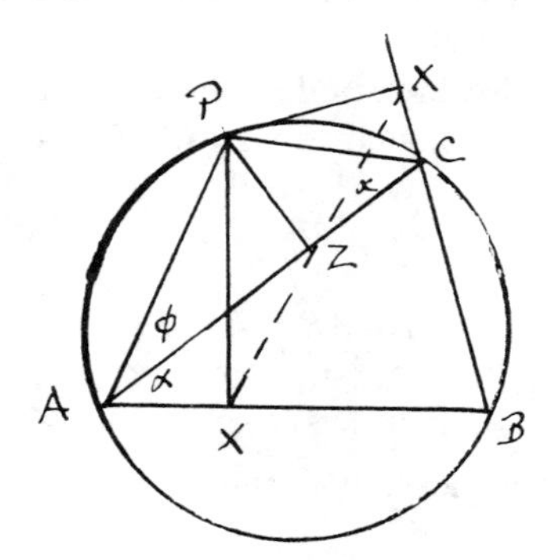

ABC is an arbitrary triangle with circumcircle shown. Take a point P on the circumcircle defined (for convenience) by the angle ϕ made by AP with AC. Let the angle between CA and AB be α. Draw perpendiculars PY, PZ to sides BC, AC, respectively. Join YZ. Join PC. Denote the angle between YZ and ZC by x. Then:

(1) CZPY is cyclic since opposite angles at Y and Z are each $\pi/2$. Hence angle YPC=x, angle YCP=$\pi/2$ - x, angle PCB=$\pi/2$ + x.

(2) CPAB is cyclic, and so $\alpha+\phi+\pi/2+x=\pi$, implying that $\alpha+\phi+x=\pi/2$. Now continue line YZ to meet AB at X. We want to show that PX is perpendicular to AB. Note that angle AZX=x, since it is opposite to angle YZC.

(3) AXZP is cyclic since angle PZX + angle PAX=$\pi/2+x+\phi+\alpha=\pi$, by (2). Therefore angle AXP=angle AZP=$\pi/2$.

Thus PX is perpendicular to AB. **QED**

This is one version of a proof that can appear in a variety of guises. The converse may be proved in similar fashion, that is: if from a point P, perpendiculars are drawn to the sides of an arbitrary triangle, then collinearity of the feet means that P lies on the circumcircle of the triangle.

As P describes the circumcircle ABC so the corresponding pedal lines generate an *envelope*, that is, a curve to which the pedal lines for each θ are tangents. To find the envelope the standard method is to write the equation of the pedal line (8.2.2) in the form $f(x,y,\theta)=0$ and then to eliminate θ between $f=0$ and $\partial f/\partial\theta=0$. The resulting equation in x and y is the envelope. Here we isolate x and y as functions of θ, thus obtaining a parametric form for each point on the envelope. This entails differentiating the equation of the line with respect to θ and using the result in the equations for x and y.

Thus

$$x=C+R\{\cos\theta+\frac{1}{2}\cos(2\sigma-2\theta)\},$$

$$y=S+R\{\sin\theta+\frac{1}{2}\sin(2\sigma-2\theta)\},$$

where

$$2C=R(\cos\alpha+\cos\beta+\cos\gamma),\quad 2S=R(\sin\alpha+\sin\beta+\sin\gamma)$$

$$2\sigma=\alpha+\beta+\gamma.$$

Figure 8.2.5

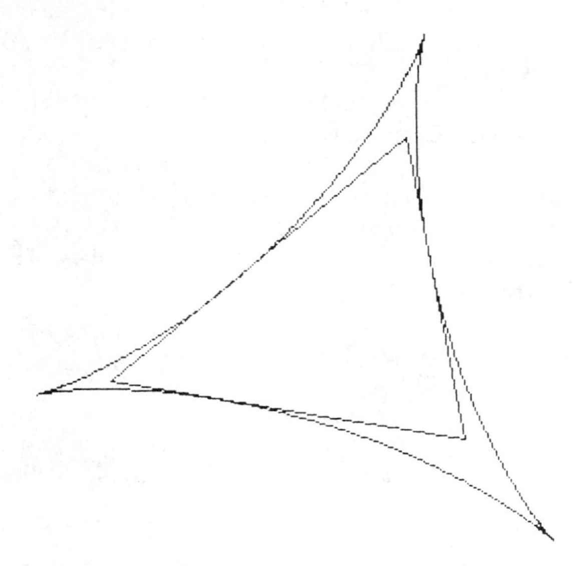

Fig. 8.2.5 shows a typical envelope. It is in the form of a star (astroid) with three double points (cusps) in the neighbourhood of the triangle's vertices. It is not surprising that the vertices should lead to double points on the envelope since from A, for example, only one non-zero length perpendicular to a side can be drawn, namely the perpendicular to BC. This is the pedal line when P is A and therefore is the tangent to the envelope at the point corresponding to $\theta=\alpha$. Note that the sides of the triangle are also tangents, for they are the pedal lines of the extremities of the diameters opposite the triangular vertices. See reference by Gale who points out that the envelope, which he identifies as a hypocycloid, is tangential to the nine-point circle of the triangle.

It is useful to know where the envelope touches the sides of the triangle. Various methods are available. One is to find the points at which the gradients of the envelope equal the slope of the sides of the triangle. The point of tangency to side AB is given by

$$x=\frac{1}{2}R\{\cos\alpha+\cos\beta-\cos\gamma+\cos(\alpha+\beta-\gamma)\}$$

$$y=\frac{1}{2}R\{\sin\alpha+\sin\beta-\sin\gamma+\sin(\alpha+\beta-\gamma)\}.$$

The three such points form the **orthic triangle**.

The cusp adjacent to vertex A has coordinates:

$$x=\frac{1}{2}R\{3\cos\alpha+\cos\beta+\cos\gamma+\cos(\beta+\gamma-\alpha)\}$$

$$y=\frac{1}{2}R\{3\sin\alpha+\sin\beta+\sin\gamma+\sin(\beta+\gamma-\alpha)\}.$$

The other cusps and points of tangency can be written down by symmetry.

8.2.2 Pedal curves

The discussion is now aimed at the more general topic of *pedal curves*. Consider as a first example the standard parabola K in the cartesian (x,y)-plane. This has equation

$$y^2=4ax,$$

axis of symmetry $y=0$, latus rectum 4a, and focus at the point (a,0). The y-axis is the tangent at the vertex. A general point T on the parabola having coordinates (x,y) can be represented in terms of a parameter t by

$$x=at^2,\quad y=2at,$$

and, as t passes from $-\infty$ to ∞, T describes the parabola. The tangent at T has slope 1/t and equation

$$ty - x - at^2 = 0.$$

Now take a general point F with coordinates (ξ,η), and draw through it the perpendicular to the tangent at T; this has slope -t and equation

$$y + tx - (\eta+t\xi) = 0.$$

It intersects the tangent at T at a point P, say, which has coordinates

$$x_P=\frac{t(\eta+\xi t)-at^2}{1+t^2},\qquad y_P=\frac{\eta+\xi t+at^3}{1+t^2}.$$

As t varies T moves along the parabola and P describes a curve k which we shall call *the pedal curve of F* with respect to the parabola K. Its equation is obtained by eliminating t and, is, in general, quite complicated. Nevertheless, it can be simple. For example, if F is the focus, $\xi=a$, $\eta=0$, k turns out to be the tangent at the vertex (here the y-axis); if F is the origin, $y_P=-tx_P$, and k is the curve

$$x(x^2+y^2)+ay^2=0.$$

This latter curve is shown in Figure 8.2.6, together with the parabola with a=1. It touches K at the origin at a double point of k.

Figure 8.2.7 shows a set of curves k corresponding to the five F-points (-2,0), (-1,0),(0,0),(1,0),(2,0). Appearances suggest that each curve lies in its own vertical strip, and this indeed turns out to be the case, the maximum width being the focal distance of F. To prove this, notice first that as $t\to\pm\infty$, $y\to\pm\infty$. Then find those t, by differentiation, which maximise/minimise x_P, and hence give $x_P(\max)-x_P(\min)$.

If F is the point $(0,\eta)$ on the y-axis we find that $tx+y=\eta$, and hence the k curves have equation

$$x^3=(\eta-y)\{x\eta-(x+a)(\eta-y)\}.$$

The pedal curve k will always touch K where a normal can be drawn to K from F. It follows that if F moves along a segment of a continuous curve of finite length, the set of pedal curves generated will envelope some part of K. This is illustrated by Figure 8.2.8 where F describes the x-axis from -5 to 5 at small intervals in ξ.

Figure 8.2.6

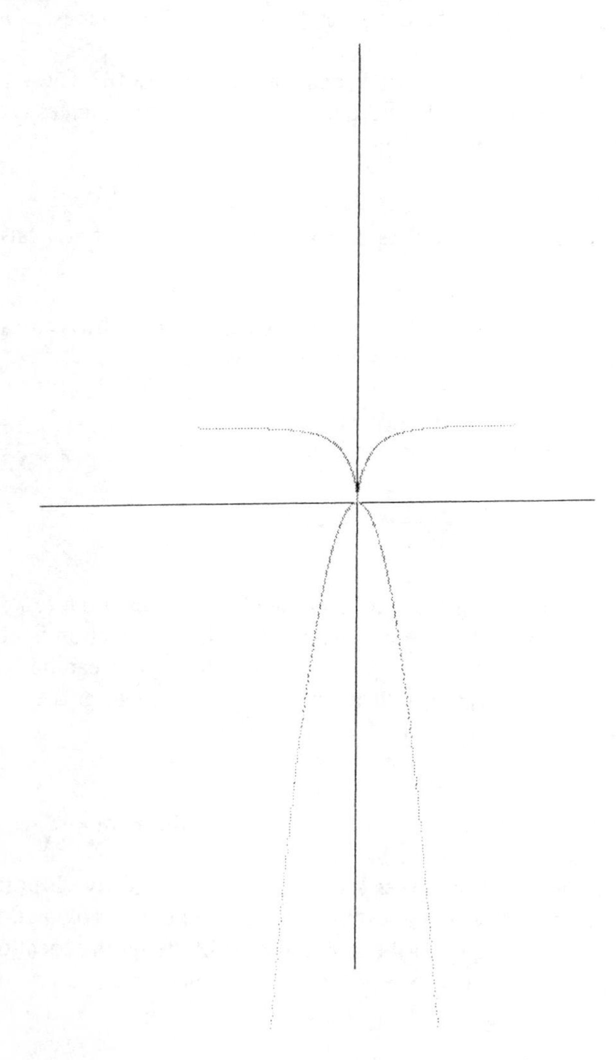

Figure 8.2.7

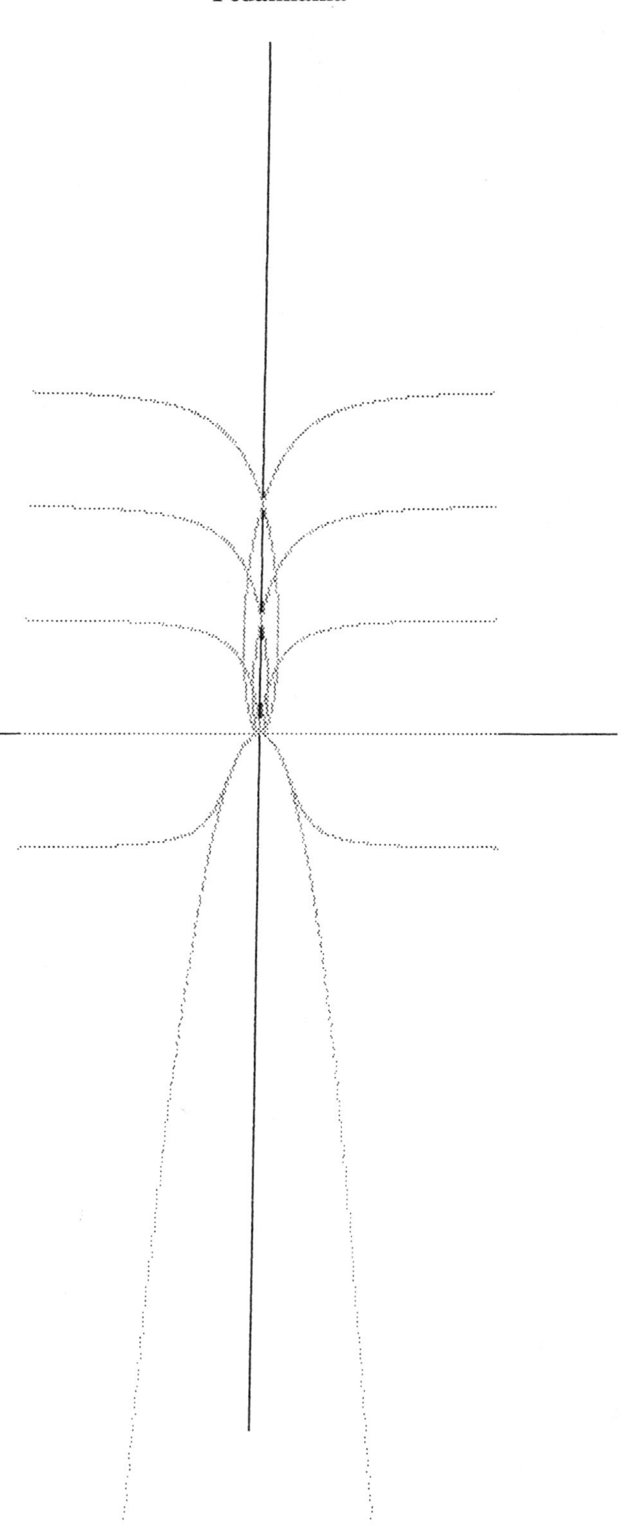

Figure 8.2.8

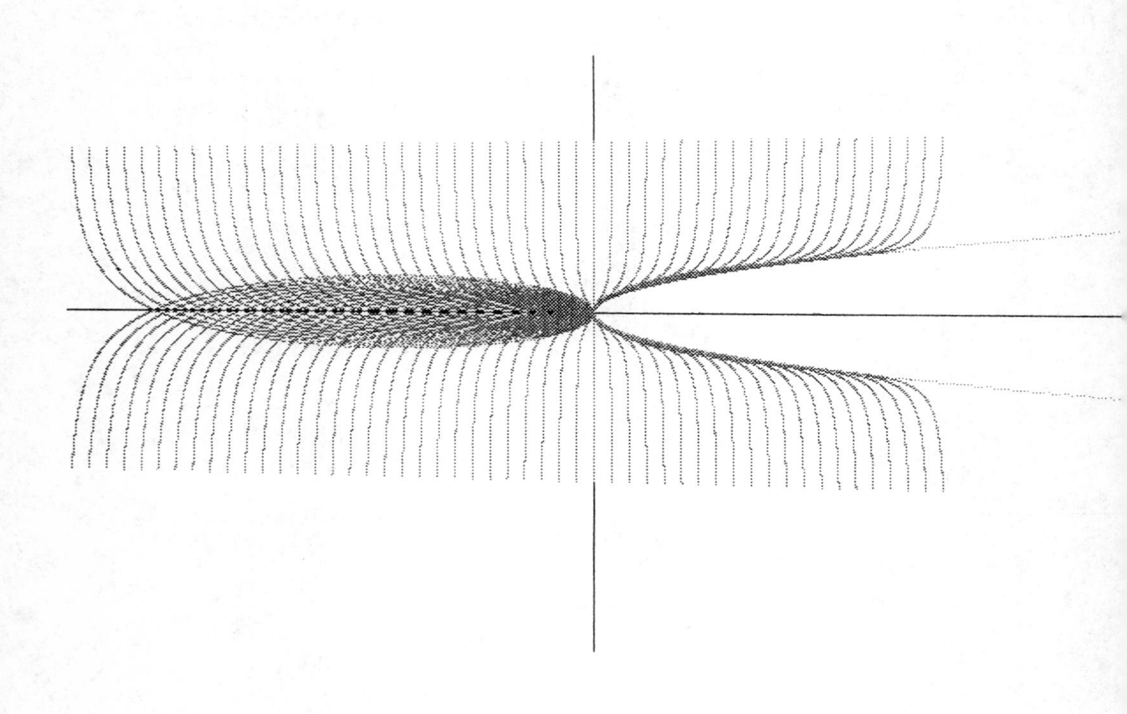

We next look at the case where K is the ellipse

$$\frac{x^2}{a^2}+\frac{y^2}{b^2}=1.$$

The centre is the origin, major and minor semiaxes are a and b respectively, along the x and y axes, and eccentricity e given by $e^2=1-b^2/a^2$. The foci have cordinates $(\pm ae,0)$. A general point on the ellipse can be written as $x=a\cos\theta$, $y=b\sin\theta$, and the ellipse is described completely as θ increases from 0 to 2π. The tangent at θ has slope $-b\cot\theta/a$ and the perpendicular to this through the point $F(\xi,\eta)$ has equation $b(y-\eta)=a\tan\theta(x-\xi)$. The
intersection of this with the tangent has coordinates

$$x_P=\frac{a(a\xi\sin^2\theta-b\eta\sin\theta\cos\theta+b^2\cos\theta)}{a^2\sin^2\theta+b^2\cos^2\theta}$$

$$y_P=\frac{b(a^2\sin\theta-a\xi\sin\theta\cos\theta+b\eta\cos^2\theta)}{a^2\sin^2\theta+b^2\cos^2\theta},$$

and is the general point P on the pedal curve k.

In the particular case where F is the origin, $y/x=a\tan\theta/b$. θ can be eliminated and the following equation for k results:

$$(x^2+y^2)^2=a^2x^2+b^2y^2.$$

This is illustrated by Figure 8.2.9.

Figure 8.2.9

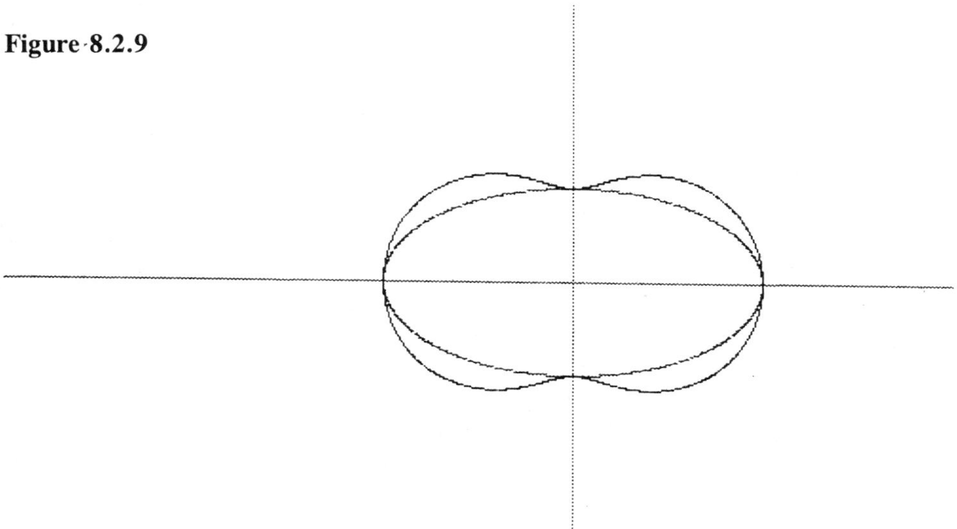

It is of interest that when F is either focus of K, k is the auxiliary circle $x^2+y^2=a^2$. This is mentioned below in connection with Hilbert's study of the inverse problem.

If K is a hyperbola the analysis is analogous. Taking the standard form

$$\frac{x^2}{a^2}-\frac{y^2}{b^2}=1,$$

with parametric point equations

$$x=a\sec\theta,\quad y=b\tan\theta,$$

so that the two branches are traced as θ increases from 0 to 2π, we get for the equations of the coordinates of a point P on the pedal curve

$$x=a\{b^2\cos\theta+(b\eta+a\xi\sin\theta)\sin\theta\}/R,$$

$$y=b\{b\eta+a\xi\sin\theta-a^2\sin\theta\cos\theta\}/R,$$

$$R=b^2+a^2\sin^2\theta.$$

When F is the origin, k is the curve

$$(x^2+y^2)^2=a^2x^2-b^2y^2.$$

In [2] Hilbert discusses the inverse problem: Given the pedal curve k and the pivot F, what is the original curve K? He tackles this by simple, pure geometry and the main result is that straight lines lead back to parabolae, while circles lead back to other conics. The proof consists in showing that the construction imposes on the points of K the focus-directrix property of points on conics.

We, however, continue in analytic vein. To get K three steps are needed:

(1) Draw a line from F to a point P on k.

(2) Draw the perpendicular PQ to FP.

(3) Find the envelope of PQ as P describes k. This is K.

For example, if k is the y-axis, F the point $(\xi,0)$, and P the point $(0,Y)$, the gradient of the line PF is $-Y/\xi$, and the equation of PQ is $Y(Y-y)+\xi x=0$. Differentiate with respect to Y and equate to zero: solve for Y. This gives $Y=y/2$. Substitution gives K, namely $y^2=4\xi x$. This is a parabola. The x-axis is axis of symmetry and the y-axis the tangent at the vertex. F is the focus. The latus rectum is 4ξ, four times the distance of F from the linear trajectory of P. This is one of the results we obtained the other way round when we proceeded from parabolic K to find k.

We can do likewise when k is a circle, confirming that K is either an ellipse or a hyperbola according as F is inside or outside the circle.

The reader may find it diverting to identify K when P describes more exotic curves k. This can provide material for budding computer artists. The Notes contain an example.

To complete 8.2.2 we prove a theorem which will throw some further light on the argument about pedal lines in 8.2.1.

Theorem 2: The circumcircle of a triangle formed by three tangents to a parabola passes through its focus.

Recall that points (x,y) on the parabola $y^2=4ax$ can be represented in parametric form by $y=2at$, $x=at^2$. The tangent at "t" is $ty-x-at^2=0$. We take three distinct tangents with $t=t_i$ (i=1,2,3) and suppose that they form a triangle ABC. The points of intersection A,B,C have coordinates

$$x_A=at_1t_2,\ y_A=a(t_1+t_2),$$

$$x_B=at_2t_3,\ y_B=a(t_2+t_3),$$

$$x_C=at_3t_1,\ y_C=a(t_3+t_1).$$

Proof Let the circumcircle have centre (ξ,η) and radius r. Its equation is

$$(x-\xi)^2+(y-\eta)^2=r^2,$$

where (x,y) take the values (x_A,y_A), etc. From the three equations so obtained we obtain easily

$$\xi=\frac{a}{2}(1+t_1t_2+t_2t_3+t_3t_1)$$

$$\eta=\frac{a}{2}(t_1+t_2+t_3-t_1t_2t_3)$$

$$r^2=\frac{a^2}{4}(1+t_3^2)\{(1-t_1t_2)^2+(t_1+t_2)^2\}.$$

The Theorem is proved by verifying that the circle passes through the focus, (a,0), i.e. by showing that $(a-\xi)^2+\eta^2=r^2$. This is routine. It follows that if the curve K is a parabola and P its focus, then the pedal curve of P with respect to K will pass through the feet of the perpendiculars from P on to the sides of the triangle, and the pedal circle will in fact be a straight line. This is the Simson Line, referred to in 8.2.1.

8.2.3 Notes on 8.2.1

Reference is made to Figure 8.2.1.

1. <u>The Gradients of the Sides of the Triangle ABC</u>

The gradient of AB is

$$(y_B-y_A)/(x_B-x_A) = (\sin\beta-\sin\alpha)/(\cos\beta-\cos\alpha),$$

easily seen to reduce to $-\cot\frac{1}{2}(\alpha+\beta)$. The other gradients can be written down directly by symmetry.

2. The Coordinates of X,Y and Z

Using the above we can write the equation of AB as

$$y\sin\tfrac{1}{2}(\alpha+\beta)+x\cos\tfrac{1}{2}(\alpha+\beta)-R\cos\tfrac{1}{2}(\alpha-\beta) = 0.$$

The perpendicular to AB through P has equation

$$y\cos\tfrac{1}{2}(\alpha+\beta)-x\sin\tfrac{1}{2}(\alpha+\beta)-r\sin\{\theta-\tfrac{1}{2}(\alpha+\beta)\} = 0.$$

The intersection is the point X and has coordinates:

$$x_X=\tfrac{1}{2}r\{\cos\theta-\cos(\alpha+\beta-\theta)\}+\tfrac{1}{2}R(\cos\alpha+\cos\beta),$$

and

$$y_X=\tfrac{1}{2}r\{\sin\theta-\sin(\alpha+\beta-\theta)\}+\tfrac{1}{2}R(\sin\alpha+\sin\beta).$$

The coordinates of Y and Z can be written down by symmetry.

3. The Simson Line

When P lies on the circumcircle, r=R and x_X and y_X can be put in the form

$$x_X=\frac{R}{2}\{\cos\alpha+\cos\beta+\cos\theta-\cos(\alpha+\beta-\theta)\},$$

$$y_X=\frac{R}{2}\{\sin\alpha+\sin\beta+\sin\theta-\sin(\alpha+\beta-\theta\}.$$

Then

$$\frac{2x_X}{R}-\cos\alpha-\cos\beta-\cos\gamma-\cos\theta=-2\cos\frac{1}{2}(\alpha+\beta+\gamma-\theta)\cos\frac{1}{2}(\alpha+\beta-\gamma-\theta),$$

$$\frac{2y_X}{R}-\sin\alpha-\sin\beta-\sin\gamma-\sin\theta=-2\sin\frac{1}{2}(\alpha+\beta+\gamma-\theta)\cos\frac{1}{2}(\alpha+\beta-\gamma-\theta).$$

From this it follows by symmetry that X,Y and Z all lie on the line

$$\{2x-R(\cos\alpha+\cos\beta+\cos\gamma+\cos\theta)\}\sin\frac{1}{2}(\alpha+\beta+\gamma-\theta)=$$

$$=\{2y-R(\sin\alpha+\sin\beta+\sin\gamma+\sin\theta)\}\cos\frac{1}{2}(\alpha+\beta+\gamma-\theta).$$

This is the Simson Line.

4. The Pedal Circle

The pedal circle with respect to P is the circumcircle of the triangle XYZ. The coordinates (ξ,η) of the centre Γ, and the radius ρ are given in the text in equations (8.2.1a). It is routine, if algebraically heavy to derive these forms: it can be done,for instance, by using the property that Γ is the intersection of the perpendicular bisectors of ZX and XY, and ρ is the distance ΓZ.

5.The Nine-point Circle

The nine-point circle of the triangle ABC passes through the mid-points of the sides, the feet of the altitudes, and the mid-points between the vertices and the orthocentre. If we write

$$2C=R(\cos\alpha+\cos\beta+\cos\gamma),\ 2S=R(\sin\alpha+\sin\beta+\sin\gamma),$$

in the coordinate system of [1] (see Figure 8.2.1), the coordinates of "interesting" points are:

(i) the mid-point of BC: $(C-\tfrac{1}{2}R\cos\alpha, S-\tfrac{1}{2}R\sin\alpha)$;

(ii) the foot of the altitude from A on to BC:

$$(C-\tfrac{1}{2}R\cos(\beta+\gamma-\alpha), S-\tfrac{1}{2}R\sin(\beta+\gamma-\alpha));$$

(iii) the orthocentre H:$(2C,2S)$;

(iv) the mid-point of HA: $(C+\tfrac{1}{2}R\cos\alpha, S+\tfrac{1}{2}R\sin\alpha)$.

It is readily seen that the equation of the nine-point circle is

$$(x-C)^2+(y-S)^2=\tfrac{1}{4}R^2.$$

The orthocentre H, the nine-point centre and circumcentre O are collinear, the line having equation

$$y = Sx/C.$$

The nine-point centre bisects HO.

Some readers may welcome further reminders about the nine-point circle. For this purpose, Figure 8.2.10 shows a triangle ABC, where D,E,F are the midpoints of the sides, and O is the circumcentre, so that OD, OE and OF are perpendicular to BC, CA, AB, respectively; L,M,N are the feet of the altitudes meeting at the orthocentre H, and U,V,W are the midpoints between the vertices and H.

Figure 8.2.10

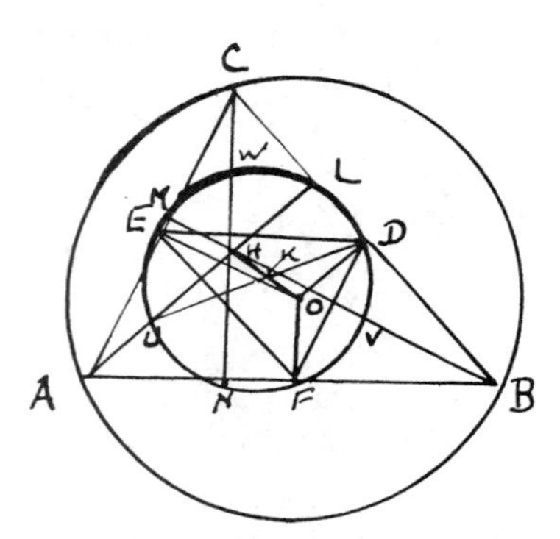

The triangle DEF is similar to ABC and has one quarter of its area. The corresponding vertices are A->D, B->E,C->F. Since O is the intersection of the altitudes of DEF, it is the orthocentre of DEF and therefore corresponds to H in ABC. Therefore, since all lengths are halved,

$$AH = 2DO, \ UH = DO,$$

and UHDO is a parallelogram with centre K, as shown.

Now draw the circle with centre K and radius KU=KD. This passes through L, because of the right-angle there. Also KD=½AO, that is half the circumradius of ABC, so the same circle passes through D as well as L and U. Starting from B (or C) instead of A, we find by similar argument that circles with the same radius pass also through E,V,M and F,W and N. These circles are the nine-point circle.

6. <u>Pedal Circles for P on AA' pass through E</u>

This is Theorem 8.2.1 in the text, where a Euclidean proof is given.If $\theta=\alpha$, (or β or γ) in (8.2.1a), it can be verified that r-R is a factor of the numerators. In consequence, the formulae assume the forms in (8.2.1c). When P coincides with A, r=R and

$$\xi = \frac{R}{4}\{3\cos\alpha + \cos\beta + \cos\gamma - \cos(\beta + \gamma - \alpha)\},$$

$$\eta = \textit{ditto with sines},$$

$$\rho = R\sin\frac{1}{2}(\alpha - \beta)\sin\frac{1}{2}(\gamma - \alpha).$$

The foot E of the altitude drawn from A has coordinates (x_E, y_E) given by (ii) above. It can then be verified that

$$(x_E - \xi)^2 + (y_E - \eta)^2 = \rho^2,$$

where ξ, η and ρ are given by (8.2.1c), so E also lies on the pedal circle.

8.2.4 Notes on 8.2.2

An example of a "more exotic" k, referred to near the end of 8.2.2, is the parabola $y^2=4ax$ whose by now familiar parametric form is $x=at^2$, $y=2at$. The slope of FP is $(2at-\eta)/(at^2-\xi)$, and the line PQ has equation

$$(2at - y)(2at - \eta) + (at^2 - x)(at^2 - \xi) = 0.$$

To find the curve K enveloped by this line as P moves round k we differentiate with respect to t, equate to zero, solve for t and, in principle, eliminate t to get K. The first stage gives the equation

$$4at - y - \eta + t(2at^2 - x - \xi) = 0.$$

Elimination of t between these looks a "tall order" in general. However, suppose that F is located on the positive x-axis, with coordinates $(\xi, 0)$. Treating the equations as simultaneous in x and y leads to the following parametric form for the coordinates of a point on K:

$$x = \frac{at^2(3at^2 + 4a - \xi)}{at^2 + \xi}$$

$$y = \frac{t\{\xi(4a - \xi) + 2a\xi t^2 - a^2 t^4\}}{at^2 + \xi}.$$

Figure 8.2.11 shows the parabola k with $a=1$ and K (like an arrow-head) when F is the origin.

Figure 8.2.11

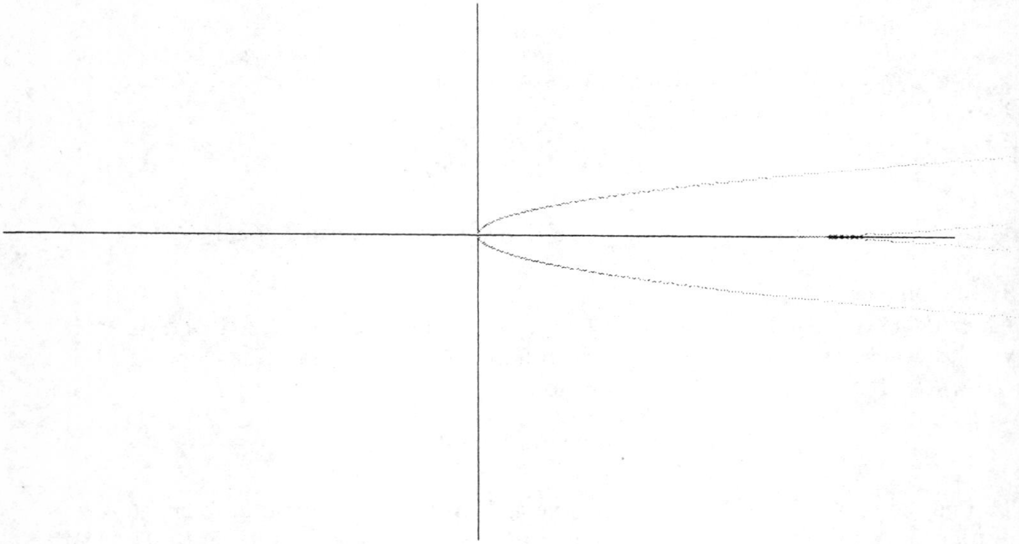

8.2.5 Note on "Napoleon's Theorem"

The equilateral triangle construction described by Figure 8.1.1 for the location of the Torricelli point of a triangle possesses other fascinating properties. One of these has been ascribed to Napoleon Buonaparte. In Figure 8.1.1 let X, Y, Z, repectively, be the centres of the equilateral triangles $A_0C_0B_1$, $B_0A_0C_1$, $C_0B_0A_1$. Napoleon's Theorem asserts that XYZ is an equilateral triangle. As illustrated by Figure 8.1.1 the three equilateral triangles on the sides of ABC are all external, but the Theorem holds good when they are drawn *internally* to the base triangle. Wetzel [7] gives an interesting discussion of aspects of the Theorem. Here is a brief trigonometric proof.

$$XC_0=b\tan\frac{\pi}{6},\quad C_0Z=a\tan\frac{\pi}{6},\quad \angle XC_0Z=\frac{\pi}{3}+\gamma,$$

$$XZ^2=XC_0^2+C_0Z^2-2XC_0.C_0Z\ \cos XC_0Z$$

$$=\tan^2\frac{\pi}{6}\left[b^2+a^2-2ab\ \cos(\frac{\pi}{3}+\gamma)\right].$$

Similarly,

$$XY^2=\tan^2\frac{\pi}{6}\left[c^2+b^2-2bc\ \cos(\frac{\pi}{3}+\alpha)\right].$$

Now express a and b in terms of c by the sine rule, i.e. $a=c\ \sin\alpha/\sin\gamma$, $b=c\ \sin\beta/\sin\gamma$. Then, to prove that XZ=XY is to prove that

$$\sin^2\alpha-2\sin\alpha\sin\beta\cos(\frac{\pi}{3}+\gamma)=\sin^2\gamma-2\sin\beta\sin\gamma\cos(\frac{\pi}{3}+\alpha).$$

This is done by replacing γ by π-$(\alpha+\beta)$. That XY=YZ follows by symmetry.

References

1.Jones, A.C. (1912) *Introduction to algebraic geometry*. Oxford.
See p.152 et seq.

2.Hilbert, D. and S.Cohn-Vossen. (1952) *Geometry and the imagination*. Chelsea, New York . This is a translation of *Anschauliche Geometrie*. Springer, Berlin, (1932), reprinted by Dover in 1944.

3.Gale, David. (1995) Triangles and computers. Mathl. Intelligencer, **17**,20-26.

4. Davis, Philip J. (1995) The rise, fall, and possible transfiguration of triangle geometry: a mini-history. American Mathematical Monthly, 102, 204-214.

5. Courant, R. and Robbins, H. (1941) *What is mathematics?* Oxford University Press.

6. Krarup, Professor Jakob, (1995). University of Copenhagen. Personal communication.

7. Wetzel, John E. (1992) Converses of Napoleon's Theorem. Amer. Math. Monthly, 99, 339.

Index

Index

Albion Publishing Series Mathematics and Applications

A MATHEMATICAL KALEIDOSCOPE

Applications in Industry, Business and Science

BRIAN CONOLLY, Emeritus Professor of Mathematics (Operational Research), University of London *and*

STEVEN VAJDA, Visiting Professor at Sussex University, *formerly* Professor of Operational Research, Department of Engineering Production, University of Birmingham

An advanced text with applications in operational research, actuarial science, engineering, communications, finance, house purchase, lotteries, gambling, management, and pursuit and search.

The wide selection of applied mathematical studies is drawn from algebra, geometry, analysis, statistics and computational methodology, each prefaced by a summary of content and mathematical relevance. Lively applicable mathematics is treated with respect and authority, with a blend of stimulus and humour, light in touch.

Contents: Miscellaneous Fantasies; Finance; Games; Mathematical Programming; Search, Pursuit, Rational Outguessing; Organisation and Management; Mathematical Teasers; Triangular Geometry.

Readership: (a) advanced undergraduates and postgraduates in applied mathematics, statistics and operational research. (b) researchers and applied mathematicians in professional practice. (c) careers advisers.

1995 276 pages 64 diagrams

ISBN: 1-898563-21-7

FUNDAMENTALS OF UNIVERSITY MATHEMATICS

COLIN McGREGOR, JOHN NIMMO and WILSON W. STOTHERS, Department of Mathematics, University of Glasgow

A unified course for first year mathematics, bridging the school/university gap, suitable for pure and applied mathematics courses, and those leading to degrees in physics, chemical physics, computing science, or statistics.

The treatment is careful, thorough and unusually clear, and the slant and terminology are modern, fresh and original, in parts sophisticated and demanding some student commitment. There are fresh ideas for teachers, students, and tutorials. There are 300 worked examples, rigorous proofs for most theorems, 750 exercises with answers are provided. Also problems and solutions for all topics covered.

Contents: Preliminaries, Functions & Inverse Functions; Polynomials & Rational Functions; Induction & the Binomial Theorem; Trigonometry, Complex Numbers; Limits & Continuity; Differentiation - Fundamentals; Differentiation - Applications; Curve Sketching; Matrices & Linear Equations; Vectors & Three Dimensional Geometry; Products of Vectors; Integration - Fundamentals; Logarithms & Exponentials; Integration - Methods & Applications; Ordinary Differential Equations; Sequences & Series; Numerical Methods.

Readership: First year undergraduates of pure and applied mathematics, physics, statistics, chemical physics and computing science.

1994 200 line drawings 540 pages

hardback ISBN: 1-898563-09-8
paperback ISBN: 1-898563-10-1

GAME THEORY: Mathematical Models of Conflict

A.J. JONES, Senior Lecturer in Computing, Imperial College of Science, Technology and Medicine, University of London

A modern, up-to-date text for senior undergraduate and graduate students (and teachers and professionals) of mathematics, economics, sociology; and operational research, psychology, defence and strategic studies, and war games. Engagingly written with agreeable humour, this account of game theory can be understood by non-mathematicians. It shows basic ideas of extensive form, pure and mixed strategies, the minimax theorem, non-cooperative and co-operative games, and a "first class" account of linear programming, theory and practice. The book is self-contained with comprehensive references from source material.

Readership: Workers, teachers and advanced students of mathematics, economics, sociology, operational research, psychology, defence and strategic studies.

1996 ISBN: 1-898563-14-4 *ca*. 300 pages

Albion Publishing Series Mathematics and Applications

CALCULUS
Introductory Theory and Applications in Physical and Life Science
R.M. JOHNSON, Department of Mathematics and Statistics, University of Paisley

This lucid and balanced text for first year undergraduates (UK) conveys the clear understanding of the fundamentals and applications of calculus, as a prelude to studying more advanced functions.
Feature: Short and fundamental diagnostic exercises at chapter ends testing comprehension, before moving to new material.
Contents: Prerequisites from Algebra, Geometry and Trigonometry; Limits and Differentiation; Differentiation of Products and Quotients; Higher-order Derivatives; Integration; Definite Integrals; Stationary Points and Points of Inflexion; Applications of the Function of a Function Rule; The Exponential, Logarithmic and Hyperbolic Functions; Methods of Integration; Further Applications of Integration; Approximate Integration; Infinite Series; Differential Equations.
Readership: First year undergraduates (including non-specialist mathematicians) of mathematics, computing, physics, engineering, chemical science, biology, and life science.

1995 ISBN: 1-898563-06-3 *ca*. 350 pages

LINEAR DIFFERENTIAL AND DIFFERENCE EQUATIONS: A Systems Approach for Mathematicians and Engineers
R.M. JOHNSON, Department of Mathematics and Statistics, University of Paisley

An advanced text for senior undergraduates and graduates, and professional workers in applied mathematics, and electrical and mechanical engineering.
Feature: The author's systematic approach lucidly explains this difficult aspect.

"Should find wide application by undergraduate students in engineering and computer science ... the author is to be congratulated on the importance that he attaches to conveying the parallelism of continuous and discrete systems"
- *Institute of Electrical Engineers (IEE) Proceedings*

1996 ISBN: 1-898563-12-5 *ca*. 200 pages

MATHEMATICS IN ELECTRONIC COMMUNICATIONS
Volume 1: NETWORKS
R.H. JONES, School of Mathematics and Information Sciences, University of Coventry

A mathematical account of important topics in communications engineering, especially aspects of design and analysis of applications in applied graph theory. A course book for advanced undergraduates and postgraduates in applied mathematics, electronics, communications and computing, also workers in industrial and academic research. There is much valuable material on operational research and discrete mathematics.
Howard Jones identifies problems, and then indicates the algorithms available for their solution. The theorem-proof approach, unpalatable to so many engineers, is conspicuous by its absence. Problem-exercises to text comprehension, with answers.

1995 ISBN: 1-898563-23-3 *ca*. 160 pages

Volume 2: SIGNAL PROCESSING
NIGEL STEELE, M. CHAPMAN and D.P. GOODALL, School of Mathematics and Information Sciences, University of Coventry

Develops the theory of communication from a mathematical viewpoint for advanced undergraduates and graduates, and professional engineers and researchers in communications engineering.
Part one focuses on continuous time models, signals and linear systems, and on system responses. Fourier methods are developed prior to a discussion of methods for the design of analogue filters. Part two discusses discrete-time signals and systems, with full development of the Z and Fourier Transforms to support the chapter on digital filter design.
Feature: An important chapter on the use of speech modelling theory provides helpful material for speech communication researchers.

Readership: Advanced undergraduates and postgraduates in applied mathematics, electronics and communications engineering, and branches of computer science. Postgraduates on higher degrees and research.

1995 ISBN: 1-898563-25-X *ca*. 280 pages